CO-BFC-922

Advances in

ACTIVATION ANALYSIS

Volume 1

Advances in
ACTIVATION ANALYSIS

Edited by

J. M. A. Lenihan
*Department of Clinical Physics and Bio-engineering,
Western Regional Hospital Board, Glasgow, Scotland*

and

S. J. Thomson
Department of Chemistry, The University, Glasgow, Scotland

Volume 1

1969

**Academic Press
London and New York**

ACADEMIC PRESS INC. (LONDON) LTD
Berkeley Square House
Berkeley Square
London, W1X 6BA

U.S. Edition published by
ACADEMIC PRESS INC.
111 Fifth Avenue
New York, New York 10003

Copyright © 1969 by ACADEMIC PRESS INC. (LONDON) LTD

All Rights Reserved
No part of this book may be reproduced in any form by photostat, microfilm, or any other means, without written permission from the publishers

Library of Congress Catalog Card Number: 75-92395
SBN: 12-000401-1

PRINTED IN GREAT BRITAIN AT
UNIVERSITY PRESS ABERDEEN

Contributors to Volume 1

H. J. M. Bowen, *Chemistry Department, The University, Reading, Berks, England.*

D. Comar, *Commissariat à l'Energie Atomique, Department de Biologie, Service Hospitalier Frédéric Joliot, Orsay, France.*

F. Girardi, *Euratom, Ispra, Italy.*

V. P. Guinn, *Technical Director, Activation Analysis Program, Gulf General Atomic Inc., San Diego, California, U.S.A.*

G. Guzzi, *Euratom, Ispra, Italy.*

W. Schulze, *Institüt für Anorganische Chemie, Freie Universität, Berlin, Germany.*

L. Sklavenitis, *Ingenieur au Commissariat à l'Energie Atomique, Centre d'Etudes Nucléaires de Saclay, France.*

R. E. Wainerdi, *Activation Analysis Research Laboratory, Department of Chemical Engineering, Texas A & M University, College Station, Texas, U.S.A.*

Preface

Activation analysis has been defined as: Instant Isotopes—or, more accurately and less frivolously, as: the revelation of chemical composition through the modification of nuclear behaviour. Most of the familiar analytical techniques involve the behaviour of electrons; mass spectrometry and nuclear magnetic resonance are among the few which invoke the properties of the nucleus. Activation analysis belongs in in this category. The versatility and sensitivity of the technique are linked with the fact that it exploits a wide range of nuclear properties, making good use of methods and instruments available as the result of progress in other parts of the nuclear realm.

Notwithstanding its relation to nuclear physics, activation analysis is essentially a chemical technique. Its relevance in biology, engineering and environmental studies derives from the remarkable sensitivity which it often provides, along with the advantages of excellent specificity, relative freedom from blank errors and, in many instances, speed and cheapness.

For a majority of the chemical elements, activation analysis will, if pushed to the limit, prove more sensitive than any other technique. It need hardly be emphasized that many factors apart from ultimate theoretical sensitivity must influence the analyst's choice. In any particular situation, it is necessary to balance the diverse and perhaps competing claims of cost, speed, ease of sample preparation and access to specialized experimental resources against sensitivity and accuracy. Nevertheless, the technique is so rich in promise—and in achievement—that none can dispute the observation of the late Richard Ogborn: Activation analysis is limited only by the imagination of the experimenter.

One purpose of the new series inaugurated by the present volume is to present authoritative studies by way of review for those practising activation analysis. Other contributions will be mainly useful by way of interpretation or exposition, as a guide to the student, teacher or research worker who wishes to inform himself on the potentialities of activation analysis in his own speciality.

The topics to be discussed will fall into two main categories. First, equipment for the production of neutrons or other agents of activation, for the measurement of induced radioactivity and for the study of other modifications of nuclear properties in irradiated samples. Radiochemistry, sample preparation and data handling procedures will also be gathered under this heading. Second, the applications of activation analysis will be illuminated.

In the present volume, Professor Schulze discusses some general principles and presents a compilation of essential data. Dr. Guinn deals with the design and operation of nuclear reactors as neutron sources. Dr. Girardi and Dr. Guzzi show how the recently-developed solid-state radiation detectors may be exploited by the activation analyst and Dr. Wainerdi describes the automation of activation analysis with the help of the electronic computer.

Three contributions deal with applications of activation analysis. Dr. Bowen reports on the development of a standard biological material and gives the results of an international exercise to compare results obtained in different laboratories. Dr. Comar examines the relevance of activation analysis in clinical science and Dr. Sklavenitis shows how activation techniques are used in the dosimetry of high-energy nuclear radiations.

Because activation analysis is such a versatile technique, its practitioners rove freely across interdisciplinary boundaries. The authors and editors associated with this series will be concerned to provide reviews which are readily comprehensible and informative to workers in adjacent fields of science and technology as well as to readers already knowledgeable on the topics under examination.

Future volumes will include contributions on forensic activation analysis, study of extra-terrestrial materials, accelerators as neutron sources, rapid radiochemical methods, well logging and other borehole techniques, applications in art and archaeology and a variety of other topics related to modern science and industry.

We shall be pleased to have the cooperation of readers in extending the scope and usefulness of the series and look forward to receiving suggestions for topics which might be dealt with in later volumes.

May, 1969

J. M. A. LENIHAN
S. J. THOMSON

Contents

Contributors to Volume 1 v

Preface vii

Activation Analysis: Some Basic Principles 1
W. SCHULZE

Reactors as Neutron Sources 37
V. P. GUINN

Automation and Electronic Data Handling.. 81
R. E. WAINERDI

Standard Materials and Intercomparisons 101
H. J. M. BOWEN

Activation Analysis Techniques for Dosimetry and Hazard Control
    with High Energy Radiations 115
L. SKLAVENITIS

γ-Ray Spectroscopy by Means of Germanium Lithium-Drifted
    Detectors in Activation Analysis 137
F. GIRARDI AND G. GUZZI

Clinical Application of Activation Analysis 163
D. COMAR

AUTHOR INDEX 207

SUBJECT INDEX 213

ACTIVATION ANALYSIS: SOME BASIC PRINCIPLES

W. SCHULZE

*Inst. f. Anorgan. Chemie, Freie Universität,
Berlin, Germany*

A. Introduction	1
B. Methods of Activation	3
1. General aspects	3
2. Activation with neutrons	4
3. Calculation of induced activity	6
4. Activation by fast neutrons	10
5. Interferences	11
6. Activation with charged particles	30
7. Activation by photons	31
C. Assay of Induced Radioactivity	31
D. Problems of Standardization	32
E. γ-Ray Spectrometry	33
F. Coincidence Methods	33
G. Limits of Detection	34
H. Comparison with other Methods of Trace Analysis	35
References	35

A. Introduction

Activation analysis is an important experimental technique arising as a by-product of the nuclear energy programmes developed during the last quarter century. It involves the production of radioactive nuclides from stable (or sometimes radioactive) elements present in the sample under investigation. Each of these radioactive nuclides has characteristic properties including the half-life and the type and energy of the emitted radiations. These characteristics can be identified by relatively simple equipment which, however, has excellent sensitivity for quantitative estimation.

Activation analysis consists of two main steps: (a) activation of the sample and (b) study of the induced radioactivity. A further stage, involving chemical treatment of the sample (for example to separate or concentrate a particular element) is sometimes added. Chemical manipulation before irradiation is hazardous, because contamination from reagents may alter the composition of the sample, particularly in regard to trace constituents.

More often, chemical treatment is interposed between irradiation and assay of induced radioactivity—for example when the resolution of the assay technique is not enough to permit the unambiguous identification and numerical estimation of the element under study. Post-irradiation chemical separation often allows the attainment of remarkable sensitivity, though not for very short-lived radionuclides. Recently developed rapid chemical separation techniques will, however, often allow effective separation of activities with half-lives of only a few seconds.

The applicability of chemical separation techniques is of course influenced by other requirements of the analyst. If the highest sensitivity and accuracy are required, a rapid answer is not essential. If, however, activation analysis is applied to the control of a chemical production process, rapid analysis is certainly necessary.

The distinctive advantages of activation analysis are as follows:
(a) The ultimate sensitivity is excellent for nearly every element and is, for many elements, better than can be obtained by any other technique.
(b) Non-destructive analysis is often possible.
(c) It is generally possible to determine several elements in a single sample, even when non-destructive methods are used.
(d) Provided that no pre-irradiation chemical manipulation is attempted, the technique is free from blank errors caused by the use of contaminated reagents.
(e) Post-irradiation chemical treatment is greatly facilitated by freedom to use carrier techniques, thus eliminating the necessity for rigorous microchemical procedures in the determination of trace constituents.
(f) In principle, activation analysis allows the opportunity of distinguishing between different isotopes of an element. This facility has been usefully exploited in a number of clinical and physiological investigations.

A number of significant limitations of activation analysis may also be mentioned.

(a) Since the technique is based on characteristics of atomic nuclei, it does not give any information about the chemical form in which a particular element is present.

(b) Near the limit of detection, the main source of error in activation analysis lies in the fluctuations associated with the statistical character of nuclear disintegrations. This statistical error is still present even when the element in question occurs in amounts well above the limit of detection. Suppose, for example, that the number of counts collected at the end of an activation analysis procedure is 100. The standard deviation will be $\sqrt{100} = \pm 10$ counts and the uncertainty in the final estimation, attributable to the random nature of the counting process, will be 10%. If the same analytical technique is applied to a more favourable sample, giving 10,000 final counts, the standard deviation will be ± 100 counts bringing an uncertainty of 1%. Thus, although the amount of material in the sample has increased 100-fold, the uncertainty in the final determination has been reduced only by a factor of 10. Using the traditional chemical methods of analysis, we should expect the experimental uncertainty to remain more nearly constant (in absolute terms) as the size of the sample is increased and therefore to diminish more rapidly as a percentage error.

It is, however, fair to add that activation analysis is only used for macro-determination when speed and convenience, rather than sensitivity, are decisive. So far, the main uses of the technique have been in microdetermination of elements close to—or beyond—the limits of sensitivity offered by other methods. Though several elements can often be determined in a single sample, it is generally not possible to obtain a complete survey of all trace constituents, as can for example be expected from spectrographic analysis.

(c) Though the chemical and instrumental methods used in activation analysis are relatively simple and inexpensive, the irradiation stage of the technique requires access to a nuclear reactor or some other source of sub-atomic particles or radiations. A considerable volume of work is necessary to justify the cost involved in the provision of such resources—though the marginal cost of providing irradiation facilities with an already established reactor or accelerator is quite small.

B. Methods of Activation

1. GENERAL ASPECTS

In principle, activation analysis may be based on any artificially-induced nuclear reaction. We may distinguish three main possibilities.

First, energy given to a nucleus by bombardment with energetic γ-ray quanta or electrons may be used to promote emission of nuclear particles or translation of nuclei into excited states.

Second, heavier particles, such as protons, deuterons and other nuclei may be accelerated by electrical means and used to bombard the sample. Such particles may be incorporated in the nucleus, altering its mass, charge and radioactive status or may bring about an internal rearrangement resulting in the emission of other nuclear particles.

Electrically-charged projectiles may be furnished with any desired energy by electrical means, but have the disadvantage that they penetrate the sample only to a short distance. Though this property can be turned to advantage in the analysis of surface layers, it is often inconvenient. Furthermore the energy of the bombarding particles is largely transformed into heat without inducing any nuclear reactions. In general, the sample must be vigorously cooled to prevent decomposition. Charged particle techniques, though characterized by serious experimental difficulties, have important applications in the activation analysis of light elements.

Third, nuclear reactions of the kind desired for activation analysis are most generally induced by neutron bombardment. Neutrons may be used to produce radioactive nuclides from almost every element, often with very high yield. In general, they have good penetration, allowing a homogeneous activation of the entire sample.

The analytical results derived from neutron activation analysis are in the form of mean values for the sample, without the disturbing effects of local irregularities. Sometimes however it is desired to investigate the distribution of a particular element over a surface. In these circumstances, spark spectroscopy, electron probe micro-analysis or charged particle activation (followed by autoradiography) are to be preferred.

2. ACTIVATION WITH NEUTRONS

The most powerful source of neutrons is the nuclear reactor, in which a chain reaction is maintained in fissile material, with the incidental production of neutrons in a broad energy spectrum. A thermal reactor contains considerable quantities of moderator (such as graphite or a hydrogenous material) in which the fission neutrons are slowed down, many of them reaching thermal energies, determined by the Boltzmann distribution appropriate to the ambient temperature. Thermal neutrons are easily captured by almost every nucleus. The thermal neutron flux inside a reactor is subject to losses (for example, by capture and escape) which can be made good only through the degradation of fast neutrons. For this reason, a steady thermal neutron flux inside a reactor is always accompanied by a substantial flux of fast neutrons, which can of course produce interfering activities. Sometimes, however, the fast neutron flux in a reactor can be put to good use since it is generally greater in magnitude and more constant in time than the neutron flux available from a 14 MeV generator. If samples are introduced in vessels containing materials (such as boron-10 and cadmium) capable of absorbing thermal neutrons, fast neutron activation analysis can be successfully accomplished in a reactor.

Though activation analysis is usually performed by exposing the analytical sample to a constant flux of neutrons for an appropriate time, an alternative method (Guinn, 1962) offers important advantages for the activation of very short-lived radionuclides. This method involves activation by a short burst of neutrons. A super-critical condition with very high neutron flux is brought about by quick removal of a control rod. The reactor is very soon shut down by mechanisms which are explained by Guinn elsewhere in the present volume (p. 73).

TABLE I

Types and Examples of Irradiation Products of (n, γ) reactions

I. Inactive isotope of mass $M+1$: $^{33}S(n, \gamma)^{34}S$

Ia. Excited (mesomeric) state of inactive isotope: $^{136}Ba(n, \gamma)^{137m}Ba \rightarrow\ ^{137}Ba$
γ

II. Mesomeric and ground state of active isotope:
$^{59}Co(n, \gamma)^{60m}Co$
$\downarrow \gamma$
$^{59}Co(n, \gamma)^{60}Co \longrightarrow\ ^{60}Ni$
β, γ

IIIa. Linear decay chain: $^{46}Ca(n, \gamma)^{47}Ca \xrightarrow{\beta}\ ^{47}Sc \xrightarrow{\beta}\ ^{47}Ti$

IIIb. Branching decay chain:
$^{76}Ge(n, \gamma)^{77m}Ge \xrightarrow{64\%}\ ^{77}As \xrightarrow{0.2\%}\ ^{77m}Se$
$\downarrow 36\% \quad 99.8\% \searrow$
$^{76}Ge(n, \gamma)^{77}Ge \nearrow \qquad\qquad\ ^{77}Se$

IV. Chain activation (secondary reaction):
$^{164}Dy(n, \gamma)^{165m}Dy$
$\qquad\qquad 2.5\% \searrow \beta, \gamma$
$97.5\% \downarrow \gamma \qquad\qquad\ ^{165}Ho$
$\qquad\qquad \beta, \gamma \nearrow$
$^{164}Dy(n, \gamma)^{165}Dy(n, \gamma)^{166}Dy \xrightarrow{\beta, \gamma}\ ^{166}Ho \xrightarrow{\beta, \gamma}\ ^{166}Er$

During the few milliseconds while the reactor is operating at an abnormally high power level, the thermal neutron flux may rise to above 10^{17} n cm^{-2} sec^{-1}—about five orders of magnitude higher than can normally be obtained from a research reactor. Pulsed reactor operation is of no advantage for the study of long-lived activities but offers improved sensitivity in investigations with short-lived nuclides.

With few exceptions, the nuclear reactions brought about by thermal neutrons are of the (n, γ) type. A reaction of this kind produces an isotope of the irradiated element with mass number increased by 1; γ-radiation is emitted momentarily as part of the nuclear rearrangement. Occasionally, radionuclides of other elements may be produced by (n, γ) reactions, as summarized in Table I. The reactions of type I are not directly interesting in activation analysis, since none of the eventual products is radioactive—but if such reactions have a large cross-section, they may diminish the neutron flux available for more fruitful reactions. This is the case for example with ^{113}Cd, from which is produced another stable isotope ^{114}Cd with an exceedingly large cross-section ($\sigma = 2{,}537$ barns) for thermal neutron capture.

3. CALCULATION OF INDUCED ACTIVITY

The formula commonly used to express the induced activity in disintegration per sec per gram of target element is based on the conception that each neutron which disappears from the incident flux is invested in activation and that the total neutron flux is only insignificantly diminished by this process. Only with these conditions is the equation

$$S = 0{\cdot}6025 \frac{\sigma \mathrm{k}}{A} \phi (1 - e^{-\lambda t_i})\ dps/g \tag{1}$$

valid for calculating the activity generated after an irradiation time of t_i in a constant flux ϕ n cm^{-2} sec^{-1}. S is the induced activity at the end of the irradiation, expressed in disintegrations per second, k is the fractional abundance of the isotope with mass number A in the target element and σ is the cross-section of the reaction, expressed in barns. The numerical factor 0·6025 has its origin in the Avogadro constant, $6{\cdot}025 \times 10^{23}$.

The measurement of the induced activity is influenced by a number of factors. Even in a non-destructive analysis, without chemical treatment, a certain waiting time, t_w must elapse before the irradiated sample arrives in the measuring system; during this time the activity is diminished by a factor $e^{-\lambda t_w}$

The relatively simple conditions of activation with thermal neutrons, attributable to the occurrence of (n, γ) reactions exclusively, sometimes become complicated because of interfering reactions. First, additional radionuclides may be deduced in decay chains; some of these nuclides may yield further radioactive species by (n, γ) reactions. Since the thermal neutron flux is almost invariably contaminated by fast neutrons, additional reactions are also possible. The cross-sections for fast neutron reactions are usually appreciable only above a threshold value

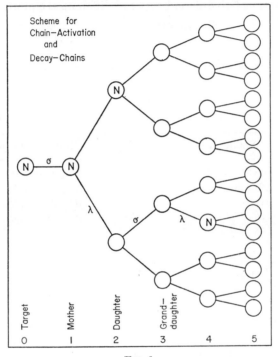

Fig. 1

of neutron energy. It is therefore necessary to know the neutron capture cross-sections as a function of neutron energy if the influence of these interfering reactions is to be estimated in advance.

Activities associated with decay chains may be studied by the following scheme, which is based on the work of Rubinson.

In Fig. 1 ascending lines of connection denote further activation reactions: descending lines denote decay. Differential equations governing the variation of the number of atoms, N, with time are generally of the form

$$\mathrm{d}N/\mathrm{d}t = \text{(production rate)} - \text{(consumption rate)}.$$

It can be seen from Fig. 1 that the production of nuclear species may take place in either of two mutually exclusive ways: (i) by decay of the predecessor characterized by a decay constant λ or (ii) by a nuclear reaction on the predecessor, characterized by a "production constant" $\phi \cdot \sigma$.

Disappearance occurs in two ways simultaneously, by decay (λ) and by nuclear reaction ($\phi\sigma$) and may be characterized by a generalized decay constant, $\Lambda = \lambda + \phi\sigma$.

There are thus two different types of differential equations:

(i) $dN_i/dt = \lambda_{i-1} N_{i-1} - \Lambda_i N_i$ and
(ii) $dN_i/dt = \phi\sigma N_{i-1} - \Lambda_i N_i$.

Table II contains the differential equations and their solutions for the

TABLE II

Differential Equations and Solutions for Chain-Activation and Decay-Chains

Definitions	Differential equations
$\Lambda = \lambda + \phi\sigma$ = generalized decay-constant	$dN_1/dt = \phi\sigma_1 N_0 - \Lambda_1 N_1$
N = number of atoms	$dN_2/dt = \lambda_1 N_1 - \Lambda_2 N_2$
$A = \lambda N$ = activity	$dN_3/dt = \lambda_2 N_2 - \Lambda_3 N_3$
	$dN_4/dt = \lambda_3 N_3 - \Lambda_4 N_4$

Solutions

$A_1 = \phi\sigma_1 N_0 \dfrac{\lambda_1}{\Lambda_1} [1 - e^{-\Lambda_1 t}]$

$A_2 = \phi\sigma_1 N_0 \dfrac{\lambda_1 \lambda_2}{\Lambda_1 \Lambda_2} \left[\dfrac{\Lambda_2}{\Lambda_1 - \Lambda_2} (1 - e^{-\Lambda_1 t}) + \dfrac{\Lambda_1}{\Lambda_1 - \Lambda_2} (1 - e^{-\Lambda_2 t}) \right]$

$A_3 = \phi\sigma_1 N_0 \dfrac{\lambda_1 \lambda_2 \lambda_3}{\Lambda_1 \Lambda_2 \Lambda_3} \left[\dfrac{\Lambda_2}{\Lambda_2 - \Lambda_1} \cdot \dfrac{\Lambda_3}{\Lambda_3 - \Lambda_1} (1 - e^{-\Lambda_1 t}) + \dfrac{\Lambda_1}{\Lambda_1 - \Lambda_2} \cdot \dfrac{\Lambda_3}{\Lambda_3 - \Lambda_2} (1 - e^{-\Lambda_2 t}) + \dfrac{\Lambda_1}{\Lambda_1 - \Lambda_3} \cdot \dfrac{\Lambda_2}{\Lambda_3 - \Lambda_3} (1 - e^{-\Lambda_3 t}) \right]$

$A_4 = \phi\sigma_1 N_0 \dfrac{\lambda_1 \lambda_2 \lambda_3 \lambda_4}{\Lambda_1 \Lambda_2 \Lambda_4 \Lambda_4} \left[\dfrac{\Lambda_2}{\Lambda_2 - \Lambda_1} \cdot \dfrac{\Lambda_3}{\Lambda_3 - \Lambda_1} \cdot \dfrac{\Lambda_4}{\Lambda_4 - \Lambda_1} (1 - e^{-\Lambda_1 t}) \right.$

$+ \dfrac{\Lambda_1}{\Lambda_1 - \Lambda_2} \cdot \dfrac{\Lambda_3}{\Lambda_3 - \Lambda_2} \cdot \dfrac{\Lambda_4}{\Lambda_4 - \Lambda_2} (1 - e^{-\Lambda_2 t})$

$+ \dfrac{\Lambda_1}{\Lambda_1 - \Lambda_3} \cdot \dfrac{\Lambda_2}{\Lambda_2 - \Lambda_3} \cdot \dfrac{\Lambda_4}{\Lambda_4 - \Lambda_3} (1 - e^{-\Lambda_3 t})$

$\left. + \dfrac{\Lambda_1}{\Lambda_1 - \Lambda_4} \cdot \dfrac{\Lambda_2}{\Lambda_2 - \Lambda_4} \cdot \dfrac{\Lambda_3}{\Lambda_3 - \Lambda_4} (1 - e^{-\Lambda_4 t}) \right]$

case of a simple decay chain of four generations. The differential equations for any other chain may be set up as in the scheme of Fig. 1. Comparison of the coefficients with those in Table II will show what adjustments are necessary to the coefficients of terms in the solution.

For example, in the case of a chain,

$$N_0 \xrightarrow{\sigma_1} N_1 \quad\quad N_3 \quad,$$
$$\lambda_1 \searrow \quad \sigma_2 \nearrow \quad \lambda_3 \searrow$$
$$N_2$$
$$\lambda_2 \searrow$$

where we are interested in the activity of N_3, the following differential equations hold:

$$dN_1/dt = \phi\sigma_1 N_0 - \lambda_1 N_1,$$
$$dN_2/dt = \lambda_1 N_1 - \Lambda_2 N_2,$$
$$dN_3/dt = \phi\sigma_2 N_2 - \lambda_3 N_3.$$

Therefore, in Table II we have to replace:

Λ_1 by λ_1, Λ_2 by $(\lambda_2 + \phi\sigma_2)$, λ_2 by $\phi\sigma_2$ and Λ_3 by λ_3.

The same replacements in the solutions given by Table II show that

$$A_3 = N_0 \frac{\phi^2 \sigma_1 \sigma_2}{\lambda_2 + \phi\sigma_2} \left[\frac{\lambda_2 + \phi\sigma_2}{\lambda_2 + \phi\sigma_2 - \lambda_1} \cdot \frac{\lambda_3}{\lambda_3 - \lambda_1} (1 - e^{-\lambda_1 t}) + \right.$$
$$\frac{\lambda_1}{\lambda_1 - \lambda_2 - \phi\sigma_2} \cdot \frac{\lambda_3}{\lambda_3 - \lambda_2 - \phi\sigma_2} \cdot (1 - e^{-(\lambda_2 + \phi\sigma_2)t}) +$$
$$\left. \frac{\lambda_1}{\lambda_1 - \lambda_3} \cdot \frac{\lambda_2 + \phi\sigma_2}{\lambda_2 + \phi\sigma_2 - \lambda_3} (1 - e^{-\lambda_3 t}) \right].$$

In branching chains the activities are found for each branch separately and added. If, at the end of the irradiation time, t_i, the activities $A_1^0, A_2^0, A_3^0, A_4^0, \ldots$ of a decay chain are present, these activities change during the waiting time, t_w, after which, for example:

$$A_4 = \lambda_2 \lambda_3 \lambda_4 A_1^0 \left[\frac{1}{(\lambda_2 - \lambda_1)(\lambda_3 - \lambda_1)(\lambda_4 - \lambda_1)} e^{-\lambda_1 t_w} + \right.$$
$$\frac{1}{(\lambda_1 - \lambda_2)(\lambda_3 - \lambda_2)(\lambda_4 - \lambda_2)} e^{-\lambda_2 t_w} +$$
$$\frac{1}{(\lambda_1 - \lambda_3)(\lambda_2 - \lambda_3)(\lambda_4 - \lambda_3)} e^{-\lambda_3 t_w} +$$
$$\left. \frac{1}{(\lambda_1 - \lambda_4)(\lambda_2 - \lambda_4)(\lambda_3 - \lambda_4)} e^{-\lambda_4 t_w} \right] +$$

$$\lambda_3\lambda_4 A_2^0 \left[\frac{1}{(\lambda_3-\lambda_2)(\lambda_4-\lambda_2)} e^{-\lambda_2 t_w} + \right.$$
$$\frac{1}{(\lambda_2-\lambda_3)(\lambda_4-\lambda_3)} e^{-\lambda_3 t_w} +$$
$$\left. \frac{1}{(\lambda_2-\lambda_4)(\lambda_3-\lambda_4)} e^{-\lambda_4 t_w} \right] +$$
$$\lambda_4 A_3^0 \left[\frac{1}{\lambda_4-\lambda_3} e^{-\lambda_3 t_w} + \frac{1}{\lambda_3-\lambda_4} e^{-\lambda_4 t_w} \right] +$$
$$A_4^0 e^{-\lambda_4 t_w}.$$

The last term of this expression, which contains A_4^0, comes from the decay of the originally present activity A_4^0 only. The last term but one, with A_3^0, describes the growth of A_4 from its precursor A_3^0. Any simpler case of genetic connection of activities may be calculated with the same formula. In the case of a parent-daughter equilibrium, A_3^0 would represent the originally present parent activity and A_4^0 the original daughter activity, while A_1^0 and A_2^0 should be put equal to zero.

By these methods it is possible to assess the potential interference—or analytical usefulness—of various activities associated with decay chains. Such assessments are, however, sometimes difficult because the cross-sections for the relevant reactions are not always known; systematic investigation of neutron capture by radionuclide is desirable in this connection.

4. ACTIVATION BY FAST NEUTRONS

Reactor neutrons are available at low cost for analytical purposes, because they occur as a by-product of the main purpose for which the reactor is used. The small space occupied by samples under irradiation need not interfere appreciably with the use of the reactor for scientific or industrial purposes. An accelerator used for the production of fast neutrons will, however, deliver a single beam of particles and nothing else. For this reason, the operating costs cannot be shared with any other function while analytical samples are being irradiated. Particle accelerators do of course have certain advantages over reactors; they are inherently safer, containing no dangerous amounts of radioactivity and no possibility of uncontrolled fission and they may be set up in any laboratory without restrictions imposed by legal considerations affecting the public safety.

In recent years significant progress has been made in the design of accelerators using the reaction ^{3}H (d, n) ^{4}He to supply neutrons of energy approximately 14 MeV at fluxes which may reach 10^9 n cm^{-2} sec^{-1}. Neutrons of this energy are appropriate for (n, p), (n, α), $(n, 2n)$ and (n, n') reactions. At 14 MeV, the cross-sections are generally considerably lower than those for (n, γ) reactions with thermal neutrons on the same target nuclei. The combination of lower flux and smaller cross-section results in diminished sensitivity for fast neutron activation analysis. The technique is, however, quite widely useful in circumstances where, for one reason or another, thermal neutron reactions are unavailable or lacking in sensitivity or specificity.

Thermal neutron irradiation may be continued for days or weeks without difficulty, since large reactors are generally operated on a 24 hr basis. Particle accelerators (including 14 MeV neutron generators) are operated intermittently, because of the limited life of the targets, and are therefore more appropriate for the production of short-lived nuclides. These difficulties have, however, provided the stimulus for the development of rapid transfer systems and quick chemical separations.

5. INTERFERENCES

The origin of a radionuclide produced by activation cannot always be determined unambiguously. If only thermal neutrons are considered, the prevalence of (n, γ) reactions simplifies the identification. With fast neutrons (or with the mixture of fast and thermal neutrons available inside a reactor) a particular product nuclide may have originated from any of several adjacent elements in the periodic table. Assessment of the significance of possible interferences is helped by reference to a table of cross-sections or, better still to a table of saturation activities and half-lives of the radionuclides concerned. With this information the activity generated under specified conditions of irradiation can be calculated by equation (1). The relevant numerical data are collected in Table III.

The production of ^{28}Al by 14 MeV neutrons may be discussed as an example of the use of this table. ^{28}Al is produced from ^{27}Al by the (n, γ) reaction, from ^{28}Si by the (n, p) reaction and from ^{31}P by the (n, γ) reaction. The saturation activities corresponding to these three reactions are:

$$^{27}\text{Al}(n,\gamma)^{28}\text{Al}: 1 \cdot 08 \times 10^4 \text{ dps/g}$$
$$^{28}\text{Si}(n,p)^{28}\text{Al}: 4 \cdot 96 \times 10^6 \text{ dps/g}$$
$$^{31}\text{P}(n,\alpha)^{28}\text{Al}: 2 \cdot 92 \times 10^6 \text{ dps/g}$$

Thermal neutrons, in the reaction ^{27}Al (n, γ) ^{28}Al give a saturation yield of $4 \cdot 69 \times 10^6$ dps/g.

TABLE III

Half-life and Saturation-activity of all Radionuclides Produced by all Stable Elements when Irradiated in a Flux of 10^9 n cm^{-2} sec^{-1} of Thermal Neutrons or of 14 MeV Neutrons.

Explanation of signs

column 1: symbol of irradiated element

column 2: type of reaction. "2" means 2n, "–" means: genetically produced; the number of identical entries denotes the number of routes by which the nuclide may be produced.

column 3: radionuclide produced, ordered by half-life.
    first group: activation by thermal neutrons
    second group: activation by 14 MeV neutrons
    Ag–110° means: there is a metastable state Ag–110 m
    Cd–109 means: no metastable state existing
    Rh–106' means: there are at least two metastable states, but without genetic relation. One dash indicates the state with the shortest half-life; two dashes—another metastable state.
    Ca–49 means: this nuclide is produced from one target element only.
    Rh–103 m means: the ground state is inactive.

column 4: ordinary half-life scale (from 0,6 sec to $1,3 \cdot 10^9$ y)

column 5: logarithm of half-life in hours

column 6: saturation activity in disintegrations per second for 1 milligram of irradiated element, containing the normal isotopic composition. (Flux of 10^9 n cm^{-2} sec^{-1}.) Lacking figures: no value available
    "–" means: genetically produced activity

column 7: "ex" means: calculation of S_∞ with an experimentally determined σ-value

    "s" means: S_∞ is calculated for each of two mesomeric states with the same σ-value. S_∞ is therefore an upper limit only.

A relatively simple method of calculating the activities produced in a sample of a particular element, allowing for the various neutron-induced reactions which may occur, has been explained by Schulze (1964).

SOME BASIC PRINCIPLES

Target		Product	$t_{\frac{1}{2}}$	log $t_{\frac{1}{2}}$ [h]	log S_∞ [dps/mg]	
H (1)	γ	H–3	12·3 y	5·03	·41–2	ex
	γ	H–3	12·3 y	5·03	·12–3	ex
He (2)	p	H–3	12·3 y	5·03	5·15	ex
	p	H–3	12·3 y	5·03	0·59	ex
	2	**He–2**				
Li (3)	γ	Li–8	·86 s	·38–4	3·46	ex
	α	H–3	12·3 y	5·03	6·84	ex
	α	**H–4**			1·91	
	p	He–7	(50 μs)		2·46	
	2n	**Li–5**	(10⁻²¹ s)		2·72	ex
	p	He–6	·82 s	·36–4	1·70	ex
	γ	Li–8	·86 s	·38–4	<·50–1	ex
	α	H–3	12·3 y	5·03	2·29	ex
Be (4)	γ	Be–10	2·5 My	10·34	2·78	ex
	α	He–6	·82 s	·36–4	2·83	ex
	γ	Be–10	2·5 My	10·34		
B (5)	p	Be–10	2·5 My	10·34	<0·35	ex
	α	Li–8	·86 s	·38–4	3·19	ex
	p	**Be–11**	14·1 s	·60–3	2·26	ex
	p	Be–10	2·5 My	10·34		
C (6)	γ	C–14	5·6 ky	7·69	·67–1	ex
	2	**C–11**	20·4 m	·52–1	0	ex
	γ	C–14	5·6 ky	7·69		
	α	Be–10	2·5 My	10·34		
N (7)	γ	N–16	7·36 s	·31–3	·55–3	ex
	p	C–14	5·6 ky	7·69	4·88	ex
	p	C–15	2·3 s	·82–4	0·88	
	γ	N–16	7·36 s	·31–3		
	2	**N–13**	10·1 m	·23–1	2·33	ex
	p	C–14	5·6 ky	7·69	3·03	
O (8)	γ	O–19	29·4 s	·91–3	·16–2	ex
	α	C–14	5·6 ky	7·69	0·49	ex
	p	**N–18**			0·20	
	α	C–15	2·3 s	·82–4	·88–1	
	p	**N–17**	4·74 s	·06–3	0·18	
	p	N–16	7·36 s	·31–3	3·13	ex
	γ	O–19	29·4 s	·91–3		
	2	**O–15**	2·05 m	·54–2	0	ex
	α	C–14	5·6 ky	7·69		
F (9)	γ	F–20	10·7 s	·48–3	2·50	ex
	α	N–16	7·36 s	·31–3	2·80	ex
	γ	F–20	10·7 s	·48–3	2·45	ex
	p	O–19	29·4 s	·91–3	3·63	ex
	2	**F–18**	1·87 h	0·27	3·28	ex

Target		Product	$t_{\frac{1}{2}}$	log $t_{\frac{1}{2}}$ [h]	log S_∞ [dps/mg]	
Ne (10)	γ	Ne–23	40·2 s	·05–2	1·94	ex
	p	**F–22**			3·04	
	p	**F–21**	5 s	·14–3	1·00	
	p	F–20	10·7 s	·48–3	4·32	
	2	**Ne–19**	19 s	·72–3	1·15	
	α	O–19	29·4 s	·91–3	1·78	
	γ	Ne–23	40·2 s	·05–2		
Na (11)	γ	Na–24	15 h	1·17	4·15	ex
	α	F–20	10·7 s	·48–3	3·67	ex
	p	Ne–23	40·2 s	·05–2	2·95	ex
	γ	Na–24	15 h	1·17	0·94	ex
	2	**Na–22**	2·6 y	4·36	2·56	ex
Mg (12)	γ	Mg–27	9·45 m	·20–1	1·85	ex
	p	**Na–26**	1·04 s	·46–4	1·81	ex
	2	**Mg–23**	11 s	·49–3	0·99	
	α	Ne–23	40·2 s	·05–2	1·96	ex
	p	**Na–25**	60 s	·22–2	2·02	ex
	γ	Mg–27	9·45 m	·20–7	·69–1	ex
	p	Na–24	15 h	1·17	3·55	ex
Al (13)	γ	Al–28	2·3 m	·58–2	3·67	ex
	2	**Al–26'**	6·6 s	·26–3	<0·56	ex
	γ	Al–28	2·3 m	·58–2	1·03	ex
	p	Mg–27	9·45 m	·20–1	3·25	ex
	α	Na–24	15 h	1·17	3·41	ex
Si (14)	γ	Si–31	2·65 h	0·41	1·84	ex
	p	**Al–30**	3·3 s	·96–4	1·85	
	2	**Si–27**	4·9 s	·14–3	1·00	
	p	Al–28	2·3 m	·58–2	3·70	ex
	p	**Al–29**	6·56 m	·04–1	1·99	ex
	α	Mg–27	9·45 m	·20–1	1·50	ex
	γ	Si–31	2·65 h	0·41	·49–1	ex
P (15)	γ	P–32	14·3 d	2·53	3·61	ex
	α	Al–28	2·3 m	·58–2	3·47	ex
	2	**P–30**	2·55 m	·67–2	2·33	ex
	p	Si–31	2·65 h	0·41	3·19	ex
	γ	P–32	14·3 d	2·53		
S (16)	γ	S–37	5·04 m	·92–2	·60–1	ex
	γ	S–35	87 d	3·32	2·29	ex
	α	**Si–33**			·60–2	
	p	**P–36**			·93–2	
	2	**S–31**	2·6 s	·86–4	0·95	
	p	P–34	12·4 s	·54–3	1·80	ex
	γ	S–37	5·04 m	·92–2		
	α	Si–31	2·65 h	0·41	1·99	ex
	p	P–32	14·3 d	2·53	3·73	ex
	p	**P–33**	24·4 d	2·77	1·69	
	γ	S–35	87 d	3·32		
	2	S–35	87 d	3·32	0·32	

Target		Product	$t_{1/2}$	log $t_{1/2}$ [h]	log S_∞ [dps/mg]		Target		Product	$t_{1/2}$	log $t_{1/2}$ [h]	log S_∞ [dps/mg]	
	γ	Cl–38m	1·0 s	·44–4	1·30	ex		α	Ar–43			·54–3	
	γ	Cl–38°	37·5 m	·79–1	3·35	ex		α	Ar–45			·66–2	
	α	P–32	14·3 d	2·53	0·02	ex		p	K–46			·89–3	
	p	S–35	87·1 d	3·32	3·32	ex		p	K–48			·15–1	
	γ	Cl–36	320 ky	9·45	5·74	ex		2	Ca–39	1·0 s	·44–4	2·07	ex
	γ	Cl–38m	1·0 s	·44–4				γ	Ca–49	8·8 m	·16–1		
	2	**Cl–34°**	1·5 s	·62–4				—	**K–45**	20 m	·52–1	—	
Cl (17)	α	P–34	12·4 s	·54–3	2·88	ex		p	K–44	22 m	·56–1	1·26	ex
	p	S–37	5·04 m	·92–2	2·00	ex		—	Sc–49	57·2 m	·98–1	—	
	2	Cl–34m	32·4 m	·73–1	2·19	ex		α	Ar–41	1·83 h	0·26	0·99	
	γ	Cl–38°	37·5 m	·79–1			Ca	p	K–42	12·5 h	1·09	0·11	ex
	α	P–32	14·3 d	2·53	3·29	ex	(20)	p	**K–43**	22·4 h	1·35	0·38	
	p	S–35	87·1 d	3·32	3·48			—	Sc–47	3·43 d	1·92	—	
	γ	Cl–36	320 ky	9·45				γ	Ca–47	4·9 d	2·07		
	2	Cl–36	320 ky	9·45	3·36			2	Ca–47	4·9 d	2·07	1·36	ex
	γ	Ar–41	1·83 h	0·26	3·90	ex		α	Ar–37	34·1 d	2·91		
	γ	Ar–37	34·1 d	2·91	2·53	ex		γ	Ca–45	164 d	3·59		
	γ	Ar–39	260 y	6·36	0·88	ex		2	Ca–45	164 d	3·59	·57–1	
	p	Cl–38m	1·0 s	·44–4	·26–1	s		α	Ar–39	260 y	6·36	1·20	
	2	**Ar–35**	1·83 s	·71–4	·45–2			γ	**Ca–41**	200 ky	9·24		
	p	**Cl–40**	1·4 m	·37–2	2·83			2	**Ca–41**	200 ky	9·24	0·62	
Ar (18)	α	S–37	5·04 m	·92–2	2·53			p	K–40	1,3·10⁹ y	13·05	3·65	ex
	p	Cl–38°	37·5 m	·79–1	·26–1	s		γ	Sc–46m	19·5 s	·74–3	5·13	ex
	γ	Ar–41	1·83 h	0·26				γ	Sc–46°	85 d	3·31	5·48	ex
	γ	Ar–37	34·1 d	2·91				γ	Sc–46m	19·5 s	·74–3		
	2	Ar–37	34·1 d	2·91	·85–2		Sc	2	**Sc–44°**	3·96 h	0·60	3·38	ex
	α	S–35	87·1 d	3·32	·11–1		(21)	α	K–42	12·5 h	1·09	2·92	ex
	γ	Ar–39	260 y	6·36				γ	**Sc–44m**	2·44 d	1·77	3·30	ex
	2	Ar–39	260 y	6·36	4·00			γ	Sc–46°	85 d	3·31		
	p	Cl–36	320 ky	9·45	1·74			p	Ca–45	164 d	3·59	2·86	ex
	γ	K–42	12·5 h	1·09	3·12	ex		γ	Ti–51	5·79 m	·98–2	1·95	ex
	γ	K–40	1,3·10⁹ y	13·05	4·44	ex		p	Sc–46m	19·5 s	·74–3	2·74	exs
	2	**K–38'**	·95 s	·42–4				p	**Sc–50°**	1·7 m	·45–2	1·23	exs
	α	Cl–38m	1·0 s	·44–4				γ	Ti–51	5·79 m	·98–2	0·34	ex
	2	**K–38"**	7·75 m	·11–1	1·86	ex		p	**Sc–50m**	23 m	·58–1	1·23	exs
K (19)	α	Cl–38°	37·5 m	·79–1	1·71	ex	Ti	p	Sc–49	57·2 m	·98–1	1·29	ex
	p	Ar–41	1·83 h	0·26	1·91	ex	(22)	2	**Ti–45**	3·1 h	0·49	1·47	ex
	γ	K–42	12·5 h	1·09	0·55	ex		p	Sc–48	1·83 d	1·64	2·92	ex
	p	Ar–39	260 y	6·36	3·59	ex		p	Sc–47	3·43 d	1·92	2·33	ex
	α	Cl–36	320 ky	9·45	3·22	ex		—	Sc–47	3·43 d	1·92	—	
	γ	K–40	1,3·10⁹ y	13·05				α	Ca–47	4·9 d	2·07	0·88	
	2	K–40	1,3·10⁹ y	13·05	1·74			p	Sc–46°	85 d	3·31	2·74	exs
								α	Ca–45	164 d	3·59	2·64	
	γ	**Ca–49**	8·8 m	·16–1	1·41	ex							
	—	Sc–49	57·2 m	·98–1	—			γ	V–52	3·76 m	·80–2	4·72	ex
Ca (20)	—	Sc–47	3·43 d	1·92	—			γ	V–52	3·76 m	·80–2	0·56	
	γ	Ca–47	4·9 d	2·07	·03–1	ex	V (23)	p	Ti–51	5·79 m	·98–2	2·50	ex
	α	Ar–47	34·1 d	2·91	1·57	ex		α	Sc–48	1·83 d	1·64	2·55	ex
	γ	Ca–45	164 d	3·59	2·31	ex		p	Sc–47	3·43 d	1·92		
	γ	**Ca–41**	200 ky	9·24	3·51	ex		2	V–49	330 d	3·90	·32–2	ex

SOME BASIC PRINCIPLES

Target		Product	$t_{\frac{1}{2}}$	$\log t_{\frac{1}{2}}$ [h]	$\log S_\infty$ [dps/mg]	
Cr (24)	γ	Cr-55	3·52 m	·77-2	1·99	ex
	γ	Cr-51	27·8 d	2·83	3·92	ex
	p	**V-54**	55 s	·78-2	0·98	
	p	**V-53**	1·7 m	·45-2	1·83	
	γ	Cr-55	3·52 m	·77-2		
	p	V-52	3·76 m	·80-2	2·89	ex
	α	T-51	5·79 m	·98-2	0·65	
	2	**Cr-49**	41·9 m	·85-1	1·11	ex
	2	Cr-51	27·8 d	2·83	3·72	
	γ	Cr-51	27·8 d	2·83		
	—	V-49	330 d	3·90	—	
Mn (25)	γ	Mn-56	2·58 h	0·41	5·16	ex
	p	Cr-55	3·52 m	·77-2	2·65	
	α	V-52	3·76 m	·80-2	2·74	ex
	γ	Mn-56	2·58 h	0·41	0·93	ex
	2	Mn-54	300 d	3·86	3·99	ex
Fe (26)	γ	Fe-59	45·1 d	3·03	1·51	ex
	γ	Fe-55	2·94 y	4·41	3·26	ex
	p	**Mn-58**	1·1 m	·26-2	0·18	
	p	**Mn-57**	1·7 m	·45-2	1·26	
	α	Cr-55	3·52 m	·77-2	·89-1	
	2	**Fe-53**	8·9 m	·17-1	1·02	ex
	p	Mn-56	2·58 h	0·41	3·11	ex
	α	Cr-51	27·8 d	2·83	2·28	
	γ	Fe-59	45·1 d	3·03		
	p	Mn-54	300 d	3·86	2·48	ex
	γ	Fe-55	2·94 y	4·41		
	2	Fe-55	2·94 y	4·41	3·69	ex
	—	**Mn-53**	140 y	6·09	—	
Co (27)	γ	Co-60m	10·5 m	·25-1	5·21	ex
	γ	Co-60°	5·3 y	4·67	5·31	ex
	γ	Co-60m	10·5 m	·25-1	0·74	s
	γ	Mn-56	2·58 h	0·41	2·49	ex
	2	Co-58m	9·0 h	0·95	3·95	sex
	p	Fe-59	45·1 d	3·03	2·72	
	2	Co-58°	7·2 d	3·24	3·95	sex
	γ	Co-60°	5·3 y	4·67	0·74	s
Ni (28)	γ	Ni-65	2·56 h	0·41	2·24	ex
	γ	Ni-63	80 y	5·84	3·73	ex
	γ	**Ni-59**	80 ky	8·84	4·47	ex
	p	Co-62′	1·6 m	·43-2	1·08	sex
	p	**Co-64′**	2·0 m	·52-2	·74-1	ex
	α	**Fe-61**	5·5 m	·96-2	0·08	
	p	**Co-64″**	7·8 m	·11-1	·34-1	ex
	p	Co-60m	10·5 m	·25-1	2·80	sex
	p	Co-62″	13·9 m	·37-7	1·08	sex
	p	**Co-61**	1·65 h	0·21	1·35	ex
	—	**Co-61**	1·65 h	0·21	—	
Ni (28)	γ	Ni-65	2·56 h	0·41		
	p	Co-58m	9 h	0·95	3·43	sex
	2	**Ni-57**	37 h	1·57	2·45	ex
	α	Fe-59	45·1 d	3·03	1·04	
	p	Co-58°	72 d	3·24	3·43	sex
	—	**Co-57**	270 d	3·81	—	
	α	Fe-55	2·94 y	4·41	3·36	
	p	Co-60°	5·3 y	4·67	2·80	sex
	γ	Ni-63	80 y	5·84		
	2	Ni-63	80 y	5·84	1·94	
	γ	**Ni-59**	80 ky	8·84		
	2	**Ni-59**	80 ky	8·84	2·88	
Cu (29)	γ	Cu-66	5·15 m	·93-2	3·76	ex
	γ	Cu-64	12·8 h	1·10	4·43	ex
	α	Co-62′	1·6 m	·43-2	1·57	s
	γ	Cu-66	5·15 m	·93-2	1·24	ex
	2	**Cu-62**	9·8 m	·21-1	3·54	ex
	α	Co-60m	10·5 m	·25-1	2·38	s
	α	Co-62″	13·9 m	·37-1	1·57	s
	p	Ni-65	2·56 h	0·41	1·74	ex
	γ	Cu-64	12·8 h	1·10	1·24	ex
	2	Cu-64	12·8 h	1·10	3·46	ex
	α	Co-60°	5·3 y	4·67	2·38	s
	p	Ni-63	80 y	5·84	2·90	ex
Zn (30)	γ	Zn-71′	2·2 m	·56-2	0·68	ex
	γ	Zn-69°	52 m	·94-1	3·27	ex
	γ	Zn-71″	3 h	0·48		
	γ	Zn-69m	13·8 h	1·13	2·20	ex
	γ	**Zn-65**	245 d	3·77	3·26	ex
	α	**Ni-67**			·51-1	
	p	**Cu-70**			·90-1	
	p	Cu-68	32 s	·95-3	1·73	
	γ	Zn-71′	2·2 m	·56-2		
	p	Cu-66	5·15 m	·93-2	2·31	ex
	2	**Zn-63**	38·3 m	·81-1	2·96	ex
	γ	Zn-69°	52 m	·94-1		
	2	Zn-69°	52 m	·94-1	1·70	s
	α	Ni-65	2·56 h	0·41	1·12	ex
	γ	Zn-71″	3 h	0·48		
	p	Cu-64	12·8 h	1·10	3·32	ex
	γ	Zn-69m	13·8 h	1·13		
	2	Zn-69m	13·8 h	1·13	1·70	s
	p	**Cu-67**	59 h	1·77	1·26	
	γ	**Zn-65**	245 d	3·77		
	2	**Zn-65**	245 d	3·77	3·20	
	α	Ni-63	80 y	5·84	2·00	
Ga (31)	γ	Ga-70	21 m	·51-1	4·42	ex
	γ	Ga-72	14·2 h	1·14	4·22	ex
	α	Cu-68	32 s	·95-3	1·38	
	p	Zn-71′	2·2 m	·56-2	1·71	s

Target		Product	$t_{\frac{1}{2}}$	log $t_{\frac{1}{2}}$ [h]	log S_∞ [dps/mg]		Target		Product	$t_{\frac{1}{2}}$	log $t_{\frac{1}{2}}$ [h]	log S_∞ [dps/mg]	
	α	Cu–66	5·15 m	·93–2	2·57	ex		γ	Se–81m	56·8 m	·98–1	2·05	ex
	γ	Ga–70	21 m	·51–1				—	Kr–**83**m	1·88 h	0·27	—	
	2	Ga–70	21 m	·51–1	3·87	ex		—	Br–83	2·33 h	0·37	—	
Ga	p	Zn–69°	52 m	·94–1	1·67	s		γ	Se–75	127 d	3·47	3·26	ex
(31)	2	**Ga–68**	68 m	0·05	3·62	ex		γ	Se–79°	60 ky	7·72		
	p	Zn–71″	3 h	0·48	1·71	s		p	As–74m	8 s	·34–3	0·88	s
	p	Zn–69m	13·8 h	1·13	2·10	ex		—	Se–77m	17·5 s	·69–3		
	γ	Ga–72	14·2 h	1·14	0·81	ex		—	Se–77m	17·5 s	·69–3	—	
	—	Se–77m	17·5 s	·69–3	—			γ	Se–77m	17·5 s	·69–3		
	γ	Ge–75m	48 s	·13–2	2·78	ex		2	Se–77m	17·5 s	·69–3	3·18	
	γ	Ge–77m	52 s	·16–2	1·26	ex		α	Ge–75m	48 s	·13–2	1·30	s
	γ	Ge–75°	82 m	0·13	3·18	ex		α	Ge–77m	52 s	·16–2	2·18	sex
	γ	Ge–77°	12 h	1·07	2·21	ex		γ	Se–83′	69 s	·28–2	·67–1	s
	—	As–77	38·8 h	1·59	—			α	**Ge–79**	⩽1 m	⩽·22–2	0·18	
	γ	Ge–71	12 d	2·45	3·77	ex		p	**As–80**	<1 m	<·22–2	1·69	
	p	**Ga–76**	32 s	·95–3	1·79			p	**As–82**	<1 m	<·22–2	0·67	
	—	Se–77m	17·5 s	·69–3	—			—	Se–79m	3·9 m	·81–2	—	
	γ	Ge–75m	48 s	·13–2				γ	Se–79m	3·9 m	·81–2		
	2	Ge–75m	48 s	·13–2				2	Se–79m	3·9 m	·81–2	3·57	s
	γ	Ge–77m	52 s	·16–2				p	As–78m	5·5 m	·96–2	1·65	s
	α	**Zn–73**	<2 m	·52–2	0·32			—	**As–79**	9 m	·17–1	—	
Ge	α	**Zn–71′**	2·2 m	·56–2	1·65	ex		γ	Se–81°	18·2 m	·48–1		
(32)	p	**Ga–74**	7·8 m	·11–1	1·78			2	Se–81°	18·2 m	·48–1		
	p	Ga–70	21 m	·55–1	2·29	ex		γ	Se–83″	25 m	·62–1	·67–1	s
	α	Zn–69°	52 m	·94–1	1·66	s		2	**Se–73′**	44 m	·87–1	1·60	
	γ	Ge–75°	82 m	0·13			Se	γ	Se–81m	56·8m	·98–1		
	2	Ge–75°	82 m	0·13	3·04	ex	(34)	2	Se–81m	56·8 m	·98–1	3·00	ex
	α	Zn–71″	3 h	0·48	1·38			α	Ge–75°	82 m	0·13	1·30	s
	p	**Ga–73**	4·85 h	0·69	1·95	ex		p	As–78°	91 m	0·18	1·65	s
	—	**Ga–73**	4·85 h	0·69	—			—	Kr–**83**m	1·88 h	0·27	—	
	γ	Ge–77°	12 h	1·07				—	Br–83	2·33 h	0·37	—	
	α	Zn–69m	13·8 h	1·13	1·66	s		—	Br–83	2·33 h	0·37	—	
	p	Ga–72	14·2 h	1·14	2·14	ex		2	**Se–73″**	7·1 h	0·85	1·60	
	—	As–77	38·8 h	1·59	—			α	Ge–77°	12 h	1·07	2·18	s
	2	**Ge–69**	39·6 h	1·60	3·47	ex		p	As–76	26·6 h	1·42	1·56	
	γ	Ge–71	12 d	2·45				γ	As–77	38·8 h	1·59	1·63	ex
	2	Ge–71	12 d	2·45	3·23			—	As–77	38·8 h	1·59	—	
	γ	As–76	26·6 h	1·42	4·52	ex		α	Ge–71	12 d	2·45	0·64	
	2	As–74m	8 s	·34–3				p	As–74°	17·5 d	2·62	0·88	s
As	p	Ge–75m	48 s	·13–2	1·75	s		—	**As–73**	76 d	3·26	—	
(33)	p	Ge–75°	82 m	0·13	1·98	ex		γ	Se–75	127 d	3·47		
	α	Ga–72	14·2 h	1·14	1·98	ex		2	Se–75	127 d	3·47	2·69	
	γ	As–76	26·6 h	1·42	1·21			—	Se–79°	60 ky	7·72	—	
	2	As–74°	17·5 d	2·62	3·64	ex		γ	Se–79°	60 ky	7·72		
	γ	Se–77m	17·5 s	·69–3	3·70	ex			Se–79°	60 ky	7·72	3·57	s
Se	γ	Se–83′	69 s	·28–2	1·53	ex		γ	Br–80°	18·5 m	·49–1	4·51	ex
(34)	γ	Se–79m	3·9 m	·81–2	2·86	ex	Br	γ	Br–80m	4·5 h	0·65	4·05	ex
	γ	Se–81°	18·2 m	·48–1	3·27	ex	(35)	γ	Br–82	35·9 n	1·55	4·08	ex
	γ	Se–83″	25 m	·62–1	0·43	ex		p	Se–79m	3·9 m	·81–2	1·97	s
								α	As–78m	5·5 m	·96–2		
								2	Br–78	⩽6 m	·97–2	3·54	ex

SOME BASIC PRINCIPLES

Target		Product	$t_{\frac{1}{2}}$	log $t_{\frac{1}{2}}$ [h]	log S_∞ [dps/mg]		Target		Product	$t_{\frac{1}{2}}$	log $t_{\frac{1}{2}}$ [h]	log S_∞ [dps/mg]	
	p	Se–81°	18·2 m	·48–1	1·96	ex		—	**Br–77°**	58 h	1·76	—	
	γ	Br–80°	18·5 m	·49–1	0·76	ex		α	Se–75	127 d	3·47	0·30	
	2	Br–80°	18·5 m	·49–1	3·21	ex		γ	Kr–85°	10·6 y	4·96		
	p	Se–81m	56·8 m	·98–1	2·07	ex	Kr	2	Kr–85°	10·6 y	4·96	3·11	s
Br	α	As–78°	91 m	0·18	2·57	ex	(36)	α	Se–79°	60 ky	7·72	1·08	s
(35)	γ	Br–80m	4·5 h	0·65	1·02	ex		γ	Kr–81°	210 ky	9·26		
	2	Br–80m	4·5 h	0·65	3·47	ex		2	Kr–81°	210 ky	9·26	2·83	s
	α	As–76	26·6 h	1·42	1·59	ex		γ	Rb–86m	1·02 m	·23–2		
	γ	Br–82	35·9 h	1·55	1·11	ex		γ	Rb–88	17·8 m	·47–1	2·36	ex
	p	Se–79°	60 ky	7·72	1·97	s		γ	Rb–86°	18·6 d	2·64	3·67	ex
	γ	Kr–81m	13 s	·56–3				γ	Rb–86m	1·02m	·23–2		
	γ	**Kr–79'**	55 s	·18–2				2	Rb–86m	1·02m	·23–2	3·21	sex
	γ	Kr–87	78 m	0·11	1·86	ex		α	Br–84'	6 m	·00–1	1·89	sex
	γ	Kr–83m	9·88 h	0·27	4·58	ex		γ	Rb–88	17·8 m	·47–1		
	γ	Kr–85m	4·4 h	0·64	2·61	ex	Rb	2	Rb–84m	23 m	·58–1	3·55	ex
	γ	**Kr–79"**	34·5 h	1·54	1·74	ex	(37)	α	Br–84"	31·8 m	·72–1	1·89	sex
	γ	Kr–85°	10·6 y	4·96	2·39	ex		p	Kr–87	78 m	0·11	1·18	
	γ	Kr–81°	210 ky	9·26				p	Kr–85m	4·4 h	0·64	1·89	s
	γ	Kr–81m	13 s	·56–3				α	Br–82	35·9 h	1·55	2·86	ex
	2	Kr–81m	13 s	·56–3	2·83	s		γ	Rb–86°	18·6 d	2·64		
	α	Se–77m	17·5 s	·69–3	0·72			2	Rb–86°	18·6 d	2·64	3·21	ex
	α	**Kr–79'**	55 s	·18–2				2	Rb–84°	33 d	2·90	3·55	sex
	2	**Kr–79'**	55 s	·18–2	2·04	s		p	Kr–85°	10·6 y	4·96	1·89	s
	p	**Br–86**	1 m	·22–2	1·04	ex		γ	**Sr–85m**	70 m	0·07	<1·60	ex
	α	Se–83'	69 s	·28–2	0·57	s		γ	**Sr–87m**	2·8 h	0·44	3·06	ex
	α	Se–79m	3·9 m	·81–2	1·08	s		γ	Sr–89	51 d	3·08	1·45	ex
	p	Br–78°	≤6 m	·99–2	0·53	s		γ	**Sr–85°**	65 d	3·18	1·74	ex
	p	Br–84'	6 m	·00–1	1·84	s		α	Kr–81m	13 s	·56–3	0·18	s
	p	Br–78m	6·4 m	·03–1	0·53	s		p	Rb–86m	1·02 m	·23–2	1·65	sex
Kr	α	Se–81°	18·2 m	·48–1	1·42	s		p	Rb–88	17·8 m	·47–1	2·23	ex
(36)	p	Br–80°	18·5 m	·49–1	1·00	s		p	Rb–84m	23 m	·58–1	0·46	s
	α	Se–83"	25 m	·62–1	0·57	s		γ	**Sr–85m**	70 m	0·07		
	p	Br–84"	31·8 m	·72–1	1·84	s		2	**Sr–85m**	70 m	0·07	2·33	ex
	α	Se–81m	56·8 m	·98–1	1·42	s		α	Kr–83m	1·88 h	0·27	1·08	
	2	**Kr–77**	1·1 h	0·05	1·18		Sr	—	Kr–83m	1·88 h	0·27	—	
	γ	Kr–87	78 m	0·11			(38)	γ	**Sr–87m**	2·8 h	0·44		
	2	Kr–83m	1·88 h	0·27	3·58			2	Sr–87m	2·8 h	0·44	3·09	ex
	—	Kr–83m	1·88 h	0·27	—			α	Kr–85m	4·4 h	0·64	2·53	ex
	—	Kr–83m	1·88 h	0·27	—			2	**Sr–83**	36 h	1·56	1·18	ex
	—	Kr–83m	1·88 h	0·27	—			p	Rb–86°	18·6 d	2·64	1·65	sex
	p	Br–83	2·33 h	0·37	1·28			p	Rb–84°	33 d	2·90	0·46	s
	—	Br–83	2·33 h	0·37	—			γ	Sr–89	51 d	3·08		
	—	Br–83	2·33 h	0·37	—			γ	**Sr–85°**	65 d	3·19		
	γ	Kr–85m	4·4 h	0·64				2	**Sr–85°**	65 d	3·19	2·29	ex
	2	Kr–85m	4·4 h	0·64	3·11	s		—	**Rb–83**	83 d	3·30	—	
	p	Br–80m	4·5 h	0·65	1·00	s		α	Kr–85°	10·6 y	4·96		
	γ	**Kr–79"**	34·5 h	1·54				α	Kr–81°	210 ky	9·26	0·18	s
	2	**Kr–79"**	34·5 h	1·54	2·04	s	Y	γ	Y–90m	3·14 h	0·50		
	p	Br–82	35·9 h	1·55	1·43		(39)	γ	**Y–90°**	64·2 h	1·81	3·93	

Target		Product	$t_{\frac{1}{2}}$	log $t_{\frac{1}{2}}$ [h]	log S_∞ [dps/mg]	
Y (39)	α	Rb–86m	1.02 m	·23–2		
	γ	Y–90m	3.14 h	0.50		
	γ	Y–90°	64.2 h	1.81	1.29	ex
	α	Rb–86°	18.6 d	2.64	2.68	ex
	p	Sr–89	51 d	3.08	2.21	
	2	Y–88	105 d	3.40	3.56	
Zr (40)	—	Nb–97m	60 s	·22–2	—	
	—	Nb–97°	74 m	0.09		
	γ	Zr–97	17 h	1.22	0.97	ex
	—	Nb–95m	84 h	1.93		
	—	Nb–95°	35 d	2.92	—	
	γ	Zr–95	63.3 d	3.18	1.93	ex
	—	Nb–93m	10 y	4.94	—	
	γ	Zr–93	1.1 My	9.99	2.45	ex
	p	Y–96	2.3 m	·58–2	0.20	
	2	Zr–90m	0.83 s	·36–4		
	—	Nb–97m	60 s	·22–2	—	
	2	Zr–89m	4.4 m	·87–2	2.44	ex
	α	Sr–93	8.2 m	·14–1	·69–1	
	p	Y–94	16.5 m	·44–1	1.73	ex
	—	Y–91m	50.3 m	·92–1		
	p	Y–91m	50.3 m	·92–1	2.13	sex
	—	Nb–97°	74 m	0.09	—	
	α	Sr–87m	2.8 h	0.44	1.01	ex
	p	Y–90m	3.14 h	0.50		
	p	Y–92	3.6 h	0.56	1.93	ex
	α	Sr–91	9.67 h	0.98	0.65	ex
	—	Y–93	10.2 h	1.02	—	
	γ	Zr–97	17 h	1.22	·85–1	ex
	p	Y–90°	64.2 h	1.81	2.94	ex
	2	Zr–89°	79 h	1.90	3.24	ex
	—	Nb–95m	84 h	1.93	—	
	—	Nb–95m	84 h	1.93	—	
	—	Nb–95°	35 d	2.92	—	
	—	Nb–95°	35 d	2.92	—	
	α	Sr–89	51 d	3.08	1.03	ex
	p	Y–91°	58 d	3.14	2.13	sex
	—	Y–91°	58 d	3.14	—	
	γ	Zr–95	63.3 d	3.18		
	2	Zr–95	63.3 d	3.18	2.30	
	—	Nb–93m	10 y	4.94	—	
	—	Nb–93m	10 y	4.94	—	
	γ	Zr–93	1.1 My	9.99		
	2	Zr–93	1.1 My	9.99	1.09	ex
	—	Zr–93	1.1 My	9.99	—	
Nb (41)	γ	Nb–94m	6.6 m	·04–1	3.85	ex
	γ	Nb–94°	20 ky	7.25		
	γ	Nb–94m	6.6 m	·04–1	1.34	s
	2	Y–90m	3.14 h	0.50	1.81	sex

Target		Product	$t_{\frac{1}{2}}$	log $t_{\frac{1}{2}}$ [h]	log S_∞ [dps/mg]	
Nb (41)	2	Nb–92′	13 h	1.11	3.31	ex
	α	Y–90°	64.2 h	1.81	1.81	ex
	2	Nb–92″	10.1 d	2.38	3.46	
	—	Nb–93m	10 y	4.94	—	
	γ	Nb–94°	20 ky	7.25	1.34	s
	p	Zr–93	1.1 My	9.99	2.15	
Mo (42)	—	Tc–101	14 m	·37–1	—	
	γ	Mo–101	14.6 m	·38–1	2.06	ex
	—	Tc–99m	6.04 h	0.78	—	
	γ	Mo–93m	6.8 h	0.83	<0.79	ex
	γ	Mo–99	67 h	1.82	2.87	ex
	γ	Mo–93°	10 ky	7.94	<2.49	ex
	—	Tc–99°	212 ky	9.27	—	
	2	Mo–91m	1.66 s	·67–4	2.29	ex
	p	Nb–97m	60 s	·22–2		
	—	Nb–97m	60 s	·22–2	—	
	p	Nb–98′	<2 m	<.52–2	1.42	s
	p	Nb–100	3 m	·70–2	0.81	
	α	Zr–89m	4.4 m	·87–2	1.00	s
	p	Nb–94m	6.6 m	·04–1	1.51	s
	—	Tc–101	14 m	·37–1		
	γ	Mo–101	14.6 m	·38–1	0.42	
	2	Mo–91°	15.5 m	·41–1	2.51	ex
	p	Nb–98″	51 m	·93–1	1.42	s
	p	Nb–97°	74 m	0.09	1.81	ex
	—	Nb–97°	74 m	0.09	—	
	—	Tc–99m	6.04 h	0.78	—	
	—	Tc–99m	6.04 h	0.78	—	
	γ	Mo–93m	6.8 h	0.83		
	2	Mo–93m	6.8 h	0.83	2.61	s
	p	Nb–92′	13 h	1.11	1.18	sex
	α	Zr–97	17 h	1.22	0.91	ex
	p	Nb–96	23.4 h	1.36	1.34	ex
	γ	Mo–99	67 h	1.82		
	2	Mo–99	67 h	1.82	3.07	ex
	α	Zr–89°	79 h	1.90	1.00	s
	—	Nb–95m	84 h	1.93	—	
	p	Nb–95m	84 h	1.93	1.61	s
	p	Nb–92″	10.1 d	2.38	1.18	sex
	—	Nb–95°	35 d	2.92	—	
	p	Nb–95°	35 d	2.92	1.61	s
	—	Nb–91m	62 d	3.17		
	α	Zr–95	63.3 d	3.18	1.00	
	—	Nb–93m	10 y	4.94	—	
	γ	Mo–93°	10 ky	7.94		
	2	Mo–93°	10 ky	7.94	2.61	s
	p	Nb–94°	20 ky	8.25	1.51	s
	—	Tc–99°	212 ky	9.27	—	
	—	Tc–99°	212 ky	9.27	—	
	α	Zr–93	1.1 My	9.99	1.11	
	—	Nb–97°	long		—	

SOME BASIC PRINCIPLES

Target		Product	$t_{\frac{1}{2}}$	log $t_{\frac{1}{2}}$ [h]	log S_∞ [dps/mg]		Target		Product	$t_{\frac{1}{2}}$	log $t_{\frac{1}{2}}$ [h]	log S_∞ [dps/mg]	
	—	Rh–105m	40 s	·04–2	—			γ	Pd–109m	4·8 m	·90–2	2·59	ex
	—	Rh–103m	54 m	·96–1	—			γ	Pd–111°	22 m	·56–1	2·13	ex
	γ	Ru–105	4·5 h	0·66	2·88	ex		—	Rh–103m	54 m	·96–1	—	
	—	Rh–105°	36·5 h	1·56	—			γ	Pd–111m	5·5 h	0·74	1·11	ex
	γ	Ru–97	2·9 d	1·84	1·86	ex		γ	Pd–109°	13·6 h	1·13	4·19	ex
	γ	Ru–103	41 d	2·99	3·43	ex		—	Ag–111°	7·5 d	2·26	—	
	—	Tc–97°	2·6 My	10·35	—			γ	Pd–103	1·7 d	2·60	2·43	ex
	p	**Tc–102′**	5 s	·14–3	1·61	s		γ	Pd–107°	7·5 My	10·81		
	p	Tc–100	16 s	·65–3	1·45			p	**Rh–110**	3·6 s	·00–3	0·81	
	—	Rh–105m	40 s	·04–2	—			p	**Rh–108**	18 s	·71–3	1·38	
	p	**Tc–102″**	4·1 m	·83–2	1·61	s		γ	Pd–107m	21·3 s	·78–3		
	p	Tc–101	14 m	·37–1	0·31	ex		2	Pd–107m	21·3 s	·78–3	3·18	s
	—	Tc–101	14 m	·37–1	—			p	**Rh–106′**	30 s	·92–3	1·60	s
	α	Mo–101	14·6 m	·38–1	0·71			—	Ag–109m	39·2 s	·04–2	—	
	p	**Tc–104**	18 m	·48–1	1·15			—	Ag–109m	39·2 s	·04–2	—	
	p	**Tc–96m**	52 m	·94–1	1·59	s		p	Rh–105m	40 s	·04–2	—	
	—	Rh–103m	54 m	·96–1	—			—	Rh–105m	40 s	·04–2	—	
	—	Rh–103m	54 m	·96–1	—			p	Rh–104°	42 s	·07–2	1·92	sex
Ru (44)	2	**Ru–95**	98 m	0·21	2·28	ex		—	Ag–111m	76 s	·32–2	—	
	γ	Ru–105	4·5 h	0·66	1·18	ex		p	Rh–104m	4·4 m	·87–2	1·92	sex
	p	Tc–99m	6·04 h	0·78	1·57	s		α	**Ru–107**	4·8 m	·90–2	0·96	ex
	—	Tc–99m	6·04 h	0·78	—			γ	Pd–109m	4·8 m	·90–2	1·35	s
	α	Mo–93m	6·8 h	0·83	1·34	s	Pd (46)	2	Pd–109m	4·8 m	·90–2		
	—	Tc–95°	20 h	1·30	—			γ	Pd–111°	22 m	·56–1	0·13	s
	—	Rh–105°	36·5 h	1·56	—			—	**Rh–107**	23 m	·58–1	—	
	α	Mo–99	67 h	1·82	1·23			—	Cd–111m	48·6 m	·91–1	—	
	γ	**Ru–97**	2·9 d	1·84				—	Rh–103m	54 m	·96–1	—	
	2	**Ru–97**	2·9 d	1·84	1·90				Rh–103m	54 m	·96–1	—	
	p	**Tc–96°**	4·3 d	2·01	1·59	s		—	Rh–103m	54 m	·96–1	—	
	γ	Ru–103	41 d	2·99				p	Rh–106″	2 h	0·30	1·60	s
	2	Ru–103	41 d	2·99	3·04			α	Ru–105	4·5 h	0·66	0·52	ex
	—	Tc–97m	91 d	3·34	—			γ	Pd–111m	5·5 h	0·74	0·13	s
	—	Tc–97m	91 d	3·34	—			2	Pd–101	8·5 h	0·93	1·57	
	α	Mo–93°	10 ky	7·94	1·34	s		γ	Pd–109°	13·6 h	1·13	1·35	s
	p	**Tc–99°**	212 ky	9·27	1·57	s		2	Pd–109°	13·6 h	1·13	3·11	ex
	—	Tc–99°	212 ky	9·27	—			p	Rh–105°	36·5 h	1·56	2·96	ex
	p	**Tc–98**	1·5 My	10·11	0·88			—	Rh–105°	36·6 h	1·56	—	
	—	**Tc–97°**	2·6 My	10·35	—			—	**Rh–101m**	4·7 d	2·05		
	—	**Tc–97°**	2·6 My	10·35	—			—	Ag–111°	7·5 d	2·26	—	
	γ	Rh–104°	42 s	·07–2	5·91	ex		γ	Pd–103	17 d	2·60		
	γ	Rh–104m	4·4 m	·87–2	4·85	ex		2	Pd–103	17 d	2·60	2·68	
	α	Tc–100	16 s	·65–3	2·55	ex		α	Ru–103	41 d	2·99	1·23	
Rh (45)	γ	Rh–104°	42 s	·07–2				p	Rh–102	210 d	3·55	0·61	
	γ	Rh–104m	4·4 m	·87–2				—	**Rh–101°**	~5 y	~4·64	—	
	—	Rh–103m	54 m	·95–1				γ	Pd–107°	7·5 My	10·81		
	p	Ru–103	41 d	2·99	2·04			2	Pd–107°	7·5 My	10·81	3·18	s
	2	Rh–102	210 d	3·55	3·68			—	Pd–107°	7·5 My	10·81	—	
Pd (46)	γ	Pd–107m	21·3 s	·78–3			Ag (47)	γ	Ag–110°	24·2 s	·83–3	5·47	ex
	—	Ag–109m	39·2 s	·04–2				γ	Ag–108°	2·4 m	·60–2	4·94	ex
	—	Ag–111m	76 s	·32–2				γ	Ag–110m	270 d	3·81	3·88	ex

Target		Product	$t_{\frac{1}{2}}$	log $t_{\frac{1}{2}}$ [h]	log S_∞ [dps/mg]		Target		Product	$t_{\frac{1}{2}}$	log $t_{\frac{1}{2}}$ [h]	log S_∞ [dps/mg]	
	γ	Ag–108m	>5 y	⩾4·64				p	Ag–106′	24 m	·60–1	0·72	sex
	p	Pd–107m	21·3 s	·77–3	1·81	s		γ	Cd–111m	48·6 m	·91–1		
	γ	Ag–110°	24·2 s	·83–3				2	Cd–111m	48·6 m	·91–1	3·08	
	α	Rh–106′	30 s	·92–3	1·43	sex		—	Cd–111m	48·6 m	·91–1	—	
	—	Ag–109m	39·2 s	·04–2	—			—	Cd–111m	48·6 m	·91–1	—	
	α	Rh–104°	42 s	·07–2	1·46	s		γ	Cd–117°	50 m	·92–1		
	γ	Ag–108°	2·4 m	·60–2	1·94	s		—	Rh–103m	54 m	·95–1	—	
Ag (47)	2	Ag–108°	2·4 m	·60–2	3·24	ex		2	Cd–105	55 m	·96–1	1·63	
	α	Rh–104m	4·4 m	·86–2	1·46	s		—	In–117°	66 m	0·04	—	
	p	Pd–109m	4·8 m	·90–2	1·53	sex		—	In–117m	1·9 h	0·27	—	
	2	Ag–106′	24 m	·60–1	3·20	ex		γ	Cd–117m	2·9 h	0·46		
	α	Rh–106″	2 h	0·30	1·43	sex		p	Ag–112	3·2 h	0·50	1·11	ex
	p	Pd–109°	13·6 h	1·12	1·53	sex		—	In–115m	4·5 h	0·65	—	
	2	Ag–106″	8·3 d	2·30	2·83	ex		—	In–115m	4·5 h	0·65	—	
	γ	Ag–110m	270 d	3·81				—	In–115m	4·5 h	0·65	—	
	α	Ag–108m	>5 y	>4·64	1·94	s		—	In–115m	4·5 h	0·65	—	
	2	Ag–108m	>5 y	>4·64				—	Ag–113″	5·3 h	0·72		
	p	Pd–107°	7·5 My	10·82	1·81	s		p	Ag–113″	5·3 h	0·72	0·66	sex
	—	Ag–109m	39·2 s	·04–2	—		Cd (48)	α	Pd–111m	5·5 h	0·74	·99–1	sex
	—	Ag–107m	44·3 s	·09–2	—			γ	Cd–107	6·7 h	0·82		
	γ	Cd–111m	48·6 m	·91–1	2·13	ex		2	Cd–107	6·7 h	0·82	1·54	
	γ	Cd–117°	50 m	·92–1				α	Pd–109°	13·6 h	1·13	0·23	sex
	—	In–117°	66 m	0·04	—			γ	Cd–115′	53 h	1·72		
	—	In–117m	1·9 h	0·27	—			2	Cd–115′	53 h	1·72	2·43	ex
	γ	Cd–117m	2·9 h	0·46	2·77	ex		p	Ag–111°	7·5 d	2·25	1·02	sex
	—	In–115m	4·5 h	0·65	—			—	Ag–111°	7·5 d	2·25	—	
	γ	Cd–115′	53 h	1·72	3·23	ex		p	Ag–106″	8·3 d	2·30	0·72	sex
	γ	Cd–115″	43 d	3·01	2·33	ex		—	Sn–117m	14 d	2·52	—	
	γ	Cd–109	1·3 y	4·05				α	Pd–103	17 d	2·61	0·49	
	γ	Cd–113m	14 y	5·09	1·59	ex		—	Ag–105	40 d	2·98	—	
Cd (48)	p	Ag–114′	5·5 s	·14–3	1·26	s		γ	Cd–115″	43 d	3·01		
	α	Pd–107m	21·3 s	·78–3	0·94	s		2	Cd–115″	43 d	3·01	2·29	ex
	p	Ag–110°	24·2 s	·83–3	1·32	s		p	Ag–110m	270 d	3·81	1·32	s
	—	Ag–109m	39·2 s	·04–2	—			γ	Cd–109	1·3 y	4·05		
	—	Ag–109m	39·2 s	·04–2	—			2	Cd–109	1·3 y	4·05	2·75	
	—	Ag–109m	39·2 s	·04–2	—			p	Ag–108m	>5 y	>4·64	0·40	s
	—	Ag–107m	44·3 s	·09–2	—			γ	Cd–113m	14 y	5·09		
	—	Ag–107m	44·3 s	·09–2	—			2	Cd–113m	14 y	5·09	3·20	
	—	Ag–107m	44·3 s	·09–2	—			α	Pd–107°	7·5 My	10·82	0·94	s
	p	Ag–113′	72 s	·30–2	0·66	sex		γ	In–116′	13 s	·56–3	5·42	ex
	—	Ag–113′	72 s	·30–2	—			γ	In–114°	72 s	·30–2	6·65	ex
	p	Ag–111m	76 s	·32–2	1·02	sex		γ	In–116″	54·2 m	·96–1	5·89	ex
	—	Ag–111m	76 s	·32–2	—		In (49)	γ	In–114m	49 d	3·07	4·10	ex
	α	Pd–113	1·4 m	·37–2	·97–1			γ	In–116′	13 s	·56–3	2·40	ex
	p	Ag–114″	2 m	·52–2		s		γ	Ag–110°	24·2 s	·83–3	0·20	s
	p	Ag–108°	2·4 m	·60–2	0·40	s		γ	In–114°	72 s	·30–2		
	p	Ag–116	2·5 m	·62–2	0·49			2	In–114°	72 s	·30–2	3·98	sex
	α	Pd–109m	4·8 m	·90–2	0·23	sex		2	In–112°	14·5 m	·38–1	2·28	s
	α	Pd–111°	23·6 m	·59–1	·99–1	sex							

SOME BASIC PRINCIPLES

Target		Product	$t_{\frac{1}{2}}$	log $t_{\frac{1}{2}}$ [h]	log S_∞ [dps/mg]		Target		Product	$t_{\frac{1}{2}}$	log $t_{\frac{1}{2}}$ [h]	log S_∞ [dps/mg]	
	2	In–112m	20·7 m	·54–1	2·28	s		—	Cd–111m	48·6 m	·91–1	—	
	γ	In–116″	54·2 m	·96–1	1·51	ex		α	Cd–117°	50 m	·92–1	0·70	s
	α	Ag–112	3·2 h	0·50	1·13	ex		p	In–116″	54·2 m	·96–1	1·23	s
	—	In–115m	4·5 h	0·65	—			p	In–117°	66 m	0·04	·85–1	s
In	p	Cd–115′	53 h	1·73	1·90	sex		—	In–117°	66 m	0·04	—	
(49)	p	Cd–115″	43 d	3·01	1·90	sex		—	**In–113m**	1·75 h	0·23	—	
	γ	In–114m	49 d	3·07	—			—	**In–113m**	1·75 h	0·23	—	
	2	In–114m	49 d	3·07	3·98	sex		p	In–117m	1·9 h	0·27	·85–1	s
	α	Ag–110m	270 d	3·81	0·20	s		—	In–117m	1·9 h	0·27	—	
	p	Cd–113m	14 y	5·09	0·56			α	Cd–117m	2·9 h	0·46	0·70	s
	γ	Sn–125′	9·5 m	·20–1	1·76	ex		p	In–115m	4·5 h	0·65	·70–1	
	γ	**Sn–113m**	27 m	·65–1				—	In–115m	4·5 h	0·65	—	
	γ	Sn–123′	40 m	·82–1	1·57	ex		γ	Sn–121′	27·5 h	1·44		
	—	**In–113m**	1·75 h	0·23	—			2	Sn–121′	27·5 h	1·44	2·46	s
	γ	Sn–121′	27·5 h	1·44	2·37	ex		—	Sn–121′	27·5 h	1·44	—	
	γ	Sn–125″	9·5 d	2·36	0·06	ex		α	Cd–115′	53 h	1·72	0·04	ex
	γ	Sn–117m	14 d	2·52	0·65	ex		—	**In–111**	2·84 d	1·83	—	
	—	Te–125m	58 d	3·14	—			γ	Sn–125″	9·5 d	2·36		
	γ	**Sn–113°**	115 d	3·44	1·82	ex		γ	Sn–117m	14 d	2·52		
	γ	Sn–123″	131 d	3·50	·37–1	ex	Sn	2	Sn–117m	14 d	2·52	3·08	
	γ	Sn–119m	275 d	3·82	1·09	ex	(50)	—	Sn–117m	14 d	2·52	—	
	—	Sb–125	2·4 y	4·32	—			—	Sn–117m	14 d	2·52	—	
	γ	Sn–121″	>5 y	4·64	0·22	ex		α	Cd–115″	43 d	3·01		
	p	**In–122**			0·15			p	In–114m	49 d	3·06	0·08	s
	p	**In–124**			0·08			—	Te–125m	58 d	3·14	—	
	α	**Cd–121**	3·5 m	·76–2	·51–1			γ	**Sn–113°**	115 d	3·44		
	p	In–118′	5·1 s	·15–3	1·26	s		2	**Sn–113°**	115 d	3·44	1·43	s
	p	In–116′	13 s	·56–3	1·23	s		γ	Sn–123″	131 d	3·50		
Sn	p	In–120′	14 s	·59–3	1·18	s		2	Sn–123″	131 d	3·50	2·61	s
(50)	—	**In–121′**	30 s	·92–3	—			γ	Sn–119m	275 d	3·82		
	—	Ag–109m	39·2 s	·04–2	—			2	Sn–119m	275 d	3·82	3·26	
	p	In–120″	55 s	·18–2	1·18	s		—	Sn–119m	275 d	3·82	—	
	p	In–114°	72 s	·30–2	0·08	s		—	Sn–119m	275 d	3·82	—	
	—	**In–119°**	2·2 m	·56–2	—			α	Cd–109	1·3 y	4·05		
	p	**In–119°**	2·2 m	·56–2	0·72	s		—	Sb–125	2·4 y	4·32	—	
	α	**Cd–119′**	3 m	·70–2	·62–1	s		γ	Sn–121″	>5 y	4·64		
	—	**In–121″**	3·1 m	·71–2	—			2	Sn–121″	>5 y	4·64	2·46	s
	p	In–118″	4·5 m	·87–2	1·26	s		—	Sn–121″	>5 y	4·64	—	
	α	**Cd–119″**	~9 m	~·17–1	·62–1	s		α	Cd–113m	14 y	5·08	0·83	
	γ	Sn–125′	9·5 m	·20–1				γ	Sb–124m²	1·3 m	·33–2	1·79	ex
	p	In–112°	14·5 m	·38–1	0·48	s		γ	Sb–122m	3·5 m	·76–2	2·73	ex
	p	**In–119m**	17·5 m	·46–1	0·72	s		γ	Sb–124m¹	21 m	·54–1	1·79	ex
	—	**In–119m**	17·5m	·46–1	—		Sb	γ	Sb–122°	2·8 d	1·83	4·23	ex
	p	In–112m	20·7 m	·54–1	0·48	s	(51)	γ	Sb–124°	60 d	3·16	3·72	ex
	γ	**Sn–113m**	27 m	·65–1				α	In–118′	5·1 s	·15–3	0·89	s
	2	**Sn–113m**	27 m	·65–1	1·43	s		α	In–120′	14 s	·59–3	0·53	s
	2	**Sn–111**	35 m	·76–1	1·86	ex		α	In–120″	~55 s	·18–2	0·53	s
	γ	Sn–123′	40 m	·82–1				γ	Sb–124m²	1·3 m	·33–2		
	2	Sn–123′	40 m	·82–1	2·61	s							
	α	Cd–111m	48·6 m	·91–1	·74–1								

Target		Product	$t_{1/2}$	log $t_{1/2}$ [h]	log S_∞ [dps/mg]	
Sb (51)	γ	Sb–122m	3.5 m	·76–2		
	2	Sb–122m	3.5 m	·76–2	3.38	s
	α	In–118″	4.5 m	·87–2	0.89	s
	2	Sb–120′	16.4 m	·44–1	3.53	ex
	γ	Sb–124m¹	21 m	·54–1		
	p	Sn–123′	40 m	·82–1	1.00	s
	p	Sn–121′	27.5 h	1.44	1.36	s
	γ	Sb–122°	2.8 d	1.83	1.66	ex
	2	Sb–122°	2.8 d	1.83	3.40	ex
	2	Sb–120″	5.8 d	2.14	3.32	ex
	γ	Sb–124°	60 d	3.16		
	p	Sn–123″	131 d	3.50	1.00	s
	p	Sn–121″	>5 y	4.64	1.36	s
Te (52)	γ	Te–131°	24.8 m	·61–1	2.55	ex
	γ	Te–129°	74 m	0.09	2.29	ex
	γ	Te–127°	9.35 h	0.97	2.85	ex
	γ	Te–131m	30 h	1.48	<1.11	ex
	—	I–131	8.05 d	2.28	—	
	—	Xe–131m	12 d	2.45	—	
	γ	Te–121°	17 d	2.60		
	γ	Te–129m	33 d	2.90	1.35	ex
	γ	Te–125m	58 d	3.14	3.13	ex
	γ	Te–123m	104 d	3.40	2.13	ex
	γ	Te–127m	105 d	3.41	1.91	ex
	γ	Te–121m	154 d	3.57		
	—	I–129	16 My	11.15	—	
	p	Sb–124m²	1.3 m	·34–2	0.40	s
	p	Sb–122m	3.5 m	·76–2	0.32	s
	p	**Sb–130′**	7 m	·07–1	1.75	s
	α	Sn–125′	9.5 m	·20–1	0.32	s
	p	**Sb–128′**	10.3 m	·24–1	0.88	s
	p	Sb–120′	16.4 m	·43–1	·08–1	s
	p	**Sb–126m**	19 m	·50–1	0.86	s
	p	Sb–124m¹	21 m	·55–1	0.40	s
	γ	Te–131°	24.8 m	·61–1		
	p	**Sb–130″**	33 m	·74–1	1.75	s
	α	Sn–123′	40 m	·82–1	0.32	s
	γ	Te–129°	74 m	0.09		
	2	Te–129°	74 m	0.09	2.98	sex
	α	**Sn–127**	1.5 h	0.17	·90–1	ex
	γ	Te–127°	9.35 h	0.97		
	2	Te–127°	9.35 h	0.97	3.08	ex
	—	Te–127°	9.35 h	0.97	—	
	p	**Sb–128″**	9.6 h	0.98	0.88	s
	2	**Te–119′**	16 h	1.20	0.59	s
	α	Sn–121′	27.5 h	1.44	·93–1	s
	γ	Te–131m	30 h	1.48		
	—	**Sb–119**	38 h	1.58	—	
	—	**Sb–119**	38 h	1.58	—	
	--	**Sb–127**	93 h	1.97	—	

Target		Product	$t_{1/2}$	log $t_{1/2}$ [h]	log S_∞ [dps/mg]	
	p	Sb–122°	2.8 d	1.82	0.32	s
	2	**Te–119″**	4.6 d	2.04	0.59	s
	p	Sb–120″	5.8 d	2.74	·08–1	s
	—	I–131	8.05 d	2.28		
	—	I–131	8.05 d	2.28	—	
	α	Sn–125″	9.5 d	2.36	0.32	s
	—	Xe–131m	12 d	2.45	—	
	—	Xe–131m	12 d	2.45	—	
	p	**Sb–126°**	12.5 d	2.47	·086	s
	α	Sn–117m	14 d	2.53	·69–2	
	γ	Te–121°	17 d	2.60		
	2	Te–121°	17 d	2.60	1.08	s
	γ	Te–129m	33 d	2.90		
	2	Te–129m	33 d	2.90	3.98	sex
	γ	Te–125m	58 d	3.14		
	2	Te–125m	58 d	3.14	3.04	
Te (52)	—	Te–125m	58 d	3.14	—	
	—	Te–125m	58 d	3.14	—	
	—	Te–125m	58 d	3.14	—	
	p	Sb–124°	60 d	3.16		s
	γ	Te–123m	104 d	3.40		
	2	Te–123m	104 d	3.40	2.38	
	γ	Te–127m	105 d	3.41		
	2	Te–127m	105 d	3.41		
	—	Te–127m	105 d	3.41	—	
	α	Sn–123″	131 d	3.49	0.32	s
	γ	Te–121m	154 d	3.57		
	2	Te–121m	154 d	3.57	1.08	s
	α	Sn–119m	275 d	3.82	·89–1	
	p	Sb–125	2.4 y	4.32	0.48	
	—	Sb–125	2.4 y	4.32	—	
	—	Sb–125	2.4 y	4.32	—	
	α	Sn–121″	>5 y	4.64	·93–1	s
	—	I–129	16 My	11.15		
I (53)	γ	I–128	24.98 m	·62–1	4.50	ex
	α	Sb–124m²	1.3 m	·33–2		
	α	Sb–124m¹	21 m	·54–1	1.93	ex
	γ	I–128	24.98 m	·62–1	1.07	ex
	p	Te–127°	9.35 h	0.97	1.76	ex
	2	I–126	13.3 d	2.50	3.76	ex
	α	Sb–124°	60 d	3.16		
	p	Te–127m	105 d	3.41		
Xe (54)	γ	**Xe–125m**	55 s	·18–2		
	γ	Xe–127m	75 s	·32–2		
	γ	Xe–137	3.8 m	·80–2	1.77	ex
	γ	Xe–135m	15.5 m	·42–1	1.97	sex
	γ	Xe–135°	9.2 h	0.96	1.97	sex
	γ	**Xe–125°**	18 h	1.27		

SOME BASIC PRINCIPLES

Target		Product	$t_{\frac{1}{2}}$	log $t_{\frac{1}{2}}$ [h]	log S_∞ [dps/mg]	
	γ	Xe-133m	2.3 d	1.74	2.65	ex
	γ	Xe-133°	5.27 d	2.10	1.25	ex
	γ	Xe-129m	8.0 d	2.28	<2.66	ex
	γ	Xe-131m	12 d	2.46	<2.98	ex
	γ	Xe-127°	36.4 d	2.94		
	—	Ba-137m	2.6 m	·63-2		
	—	I-125	60 d	3.16		
	—	Cs-137	30 y	5.42		
	γ	Xe-125m	55 s	·18-2		
	2	Xe-125m	55 s	·18-2	0.60	s
	γ	Xe-127m	75 s	·32-2		
	2	Xe-127m	75 s	·32-2	1.97	s
	p	I-136	86 s	·38-2	0.08	
	α	**Te-133°**	2 m	·52-2	·49-1	s
	—	Ba-137m	2.6 m	·63-2		
	γ	**Xe-137**	3.8 m	·80-2		
	γ	Xe-135m	15.6 m	·42-1		
	2	Xe-135m	15.6 m	·42-1	2.79	s
	α	Te-131°	24.8 m	·61-1	·75-1	s
	p	I-128	24.98 m	·62-1	0.08	
	p	**I-134**	52.5 m	·93-1	0.28	
	α	**Te-133m**	63 m	0.02	·49-1	s
	α	Te-129°	74 m	0.09	0.34	s
	2	**Xe-123**	1.8 h	0.25	0.59	
Xe	p	**I-132**	2.33 h	0.36	0.87	
(54)	γ	Xe-135°	9.2 h	0.96		
	2	Xe-135°	9.2 h	0.96	2.79	s
	α	Te-127°	9.35 h	0.97	·74-1	s
	p	I-130	12.5 h	1.09	0.23	
	—	**I-123**	13 h	1.11		
	γ	**Xe-125°**	18 h	1.27		
	2	**Xe-125°**	18 h	1.27	0.60	s
	—	**I-133**	20.8 h	1.31		
	α	Te-131m	30 h	1.48	·75-1	
	γ	Xe-133m	2.3 d	1.74		
	2	Xe-133m	2.3 d	1.74	2.82	s
	—	Xe-133m	2.3 d	1.74		
	p	**I-124**	4.5 d	2.03	·15-1	s
	γ	Xe-133°	5.27 d	2.10		
	2	Xe-133°	5.27 d	2.10	2.82	s
	—	Xe-133°	5.27 d	2.10		
	γ	Xe-129m	8.0 d	2.28		
	2	Xe-129m	8.0 d	2.28	2.34	
	p	I-131	8.05 d	2.28	0.83	
	—	I-131	8.05 d	2.28		
	γ	Xe-131m	12 d	2.46		
	2	Xe-131m	12 d	2.46	2.20	
	—	Xe-131m	12 d	2.46		
	—	Xe-131m	12 d	2.46		
	p	**I-126**	13.3 d	2.50	·93-2	

Target		Product	$t_{\frac{1}{2}}$	log $t_{\frac{1}{2}}$ [h]	log S_∞ [dps/mg]	
	α	Te-121°	17 d	2.60	·79-2	s
	α	Te-129m	33 d	2.90	0.34	s
	γ	Xe-127°	36.4 d	2.94		
	2	Xe-127°	36.4 d	2.94	1.97	s
	α	Te-125m	58 d	3.14	·62-1	
Xe	—	**I-125**	60 d	3.16	—	
(54)	—	**I-125**	60 d	3.16	—	
	α	Te-**123m**	104 d	3.40	·53-2	
	α	Te-127m	105 d	3.40	·74-1	s
	α	Te-121m	154 d	3.57	·79-2	s
	—	Cs-137	30 y	5.42	—	
	p	I-129	16 My	11.14	1.15	
	—	I-129	16 My	11.14	—	
	γ	Cs-134m	3.2 h	0.50	4.13	ex
	γ	Cs-134°	2.3 y	4.30	5.07	ex
	γ	Cs-134m	3.2 h	0.50		
Cs	α	I-130	12.5 h	1.09	0.83	ex
(55)	p	Xe-133m	2.3 d	1.74	1.36	s
	p	Xe-133°	5.27 d	2.10	1.36	s
	2	Cs-132	6.5 d	2.19	3.72	
	γ	Cs-134°	2.3 y	4.30		
	γ	Ba137m	2.6 m	·63-2	2.14	ex
	γ	Ba-139	85 m	0.15	3.27	ex
	γ	Ba-135m	28.7 h	1.46	~2.34	ex
	γ	Ba-133m	38.8 h	1.59		
	—	Cs-131	10 d	2.38	—	
	γ	Ba-131	11.6 d	2.45	1.67	ex
	γ	Ba-133°	7.2 y	4.80	1.49	ex
	α	Xe-127m	75 s	·32-2	·62-2	s
	γ	Ba-137m	2.6 m	·63-2		
	2	Ba-137m	2.6 m	·63-2	3.62	
	—	Ba-137m	2.6 m	·63-2		
	α	Xe-135m	15.6 m	·42-1	0.67	s
Ba	p	**Cs-130**	30 m	·70-1	·04-2	
(56)	p	**Cs-138**	32 m	·72-1	0.84	ex
	γ	Ba-139	85 m	0.15	·61-1	ex
	2	**Ba-129**	2.45 h	0.39	0.62	
	p	Cs-134m	3.2 h	0.50	0.08	s
	α	Xe-135°	9.2 h	0.96	0.67	s
	γ	Ba-135m	28.7 h	1.46		
	2	Ba-135m	28.7 h	1.46	2.62	
	—	**Cs-129**	31 h	1.49	—	
	γ	Ba-133m	38.8 h	1.59		
	2	Ba-133m	38.8 h	1.59	2.08	s
	α	Xe-133m	2.3 d	1.74	·88-1	s
	α	Xe-133°	5.27 d	2.70	·88-1	s
	p	Cs-132	6.5 d	2.19	·85-2	
	α	Xe-129m	8 d	2.28	·40-2	

Target		Product	$t_{\frac{1}{2}}$	$\log t_{\frac{1}{2}}$ [h]	$\log S_\infty$ [dps/mg]		Target		Product	$t_{\frac{1}{2}}$	$\log t_{\frac{1}{2}}$ [h]	$\log S_\infty$ [dps/mg]	
	—	**Cs–131**	10 d	2·38	—		Ce	—	La–137	60 ky	8·72	—	
	—	**Cs–131**	10 d	2·38	—		(58)	—	La–137	60 ky	8·72	—	
	γ	**Ba–131**	11·6 d	2·45	—			γ	Pr–142	19·3 h	1·28	4·67	ex
	2	**Ba–131**	11·6 d	2·45	0·64		Pr	2	**Pr–140**	3·4 m	·75–2	3·93	ex
	α	Xe–131m	12 d	2·46	·59–1		(59)	γ	Pr–142	19·3 h	1·28	1·15	ex
Ba	p	Cs–136	13 d	2·49	1·12	ex		p	Ce–141	32·5 d	2·84	1·48	
(56)	α	Xe–127°	36·4 d	2·94	·62–2	s		γ	Nd–151	12 m	·30–1	2·53	ex
	p	Cs–134°	2·3 y	4·30	0·08	s		γ	Nd–149	2 h	0·30	2·87	ex
	γ	Ba–133°	7·2 y	4·80	—			—	Pm–151	27·5 h	1·44	—	
	2	Ba–133°	7·2 y	4·80	2·08	s		—	Pm–149	53·1 h	1·73	—	
	p	Cs–137	30 y	5·42	0·48			γ	Nd–147	11·3 d	2·44	3·84	ex
	p	Cs–135	2 My	10·24	0·42			—	Pm–147	2·52 y	4·34	—	
	γ	La–140	40·2 h	1·60	4·55	ex		—	Sm–151	100 y	5·94	—	
	p	Ba–139	85 m	0·15	1·41	ex		p	Pr–150			·92–1	
La	γ	La–140	40·2 h	1·60	0·81	ex		α	Ce–139m	55 s	·18–2	0·78	s
(57)	α	Cs–136	13 d	2·49	0·94	ex		2	Nd–141m	61 s	·23–2	3·08	s
	2	La–137	60 ky	8·72	—			α	**Ce–147**	1·2 m	·30–2	·36–1	
	α	Cs–135	2 My	10·24	—			p	**Pr–148**	2 m	·52–2	0·08	
	γ	Ce–139m	55 s	·18–2	·85–2	ex		α	**Ce–145**	3 m	·70–2		
	γ	**Ce–137°**	8·8 h	0·94	1·73	ex		—	Pr–147	12 m	·30–1	—	
	γ	Ce–143	33 h	1·51	2·65	ex		γ	Nd–151	12 m	·30–1		
	γ	**Ce–137m**	34·5 h	1·54	0·71	ex		p	**Pr–144**	17·5 m	·46–1	1·00	
	—	Pr–143	13·7 d	2·51				p	**Pr–146**	24·4 m	·61–1	0·70	
	γ	Ce–141	32 d	2·88	3·07	ex		γ	Nd–149	2 h	0·30		
	γ	Ce–139°	140 d	3·52	~1·04	ex	Nd	2	Nd–149	2 h	0·30	2·52	
	γ	Ce–139m	55 s	·18–2			(60)	2	Nd–141°	2·42 h	0·38	3·08	s
	2	Ce–139m	55 s	·18–2	3·65	s		p	**Pr–145**	6 h	0·78	0·49	
	α	Ba–137m	2·6 m	·63–2	1·66	ex		—	**Pr–145**	6 h	0·78	—	
	p	**La–136**	9·5 m	·20–1	·20–1			p	Pr–142	19·3 h	1·28	1·17	ex
	p	**La–142**	81 m	0·13	1·63	ex		—	Pm–151	27·5 h	1·44	—	
	α	Ba–139	85 m	0·15	1·52	ex		α	Ce–143	33 h	1·51	0·33	ex
	γ	**Ce–137°**	8·8 h	0·94				—	Pm–149	53·1 h	1·73	—	
	2	**Ce–137°**	8·8 h	0·94	1·08	s		—	Pm–149	53·1 h	1·73	—	
Ce	—	La–135	19 h	1·27	—			γ	Nd–147	11·3 d	2·44		
(58)	2	**Ce–135**	22 h	1·34	0·91			2	Nd–147	11·3 d	2·44	2·51	
	α	Ba–135m	28·7 h	1·46	·67–2			—	Nd–147	11·3 d	2·44	—	
	γ	Ce–143	33 h	1·51	0·52	ex		p	Pr–143	13·7 d	2·51	0·79	ex
	γ	**Ce–137m**	34·5 h	1·54				—	Pr–143	13·7 d	2·51	—	
	2	**Ce–137m**	34·5 h	1·54	1·08	s		α	Ce–141	32·5 d	2·89	0·53	
	α	Ba–133m	38·8 h	1·59	·78–2	s		α	Ce–139°	140 d	3·52	0·78	s
	p	La–140	40·2 h	1·60	1·66	ex		—	Pm–147	2·52 y	4·34	—	
	—	Pr–143	13·7 d	2·51	—			—	Pm–147	2·52 y	4·34	—	
	γ	Ce–141	32 d	2·88				—	Pm–147	2·52 y	4·34	—	
	2	Ce–141	32 d	2·88	2·78			—	Sm–151	100 y	5·94	—	
	α	Ce–139°	140 d	3·52				γ	Sm–155	24 m	·60–1	3·68	ex
	2	Ce–139°	140 d	3·52	3·65	s	Sm	γ	Sm–153	47 h	1·67	5·17	ex
	α	Ba–133°	7·2 y	4·80	·78–2	s	(62)	γ	Sm–145	360 d	3·98	<2·42	ex
	—	La–137	60 ky	8·72	—			—	Eu–155	1·7 y	4·17	—	
	—	La–137	60 ky	8·72	—			—	**Pm–145**	18 y	5·20	—	

SOME BASIC PRINCIPLES

Target		Product	$t_{\frac{1}{2}}$	log $t_{\frac{1}{2}}$ [h]	log S_∞ [dps/mg]	
	γ	Sm–151	~100 y	5·94		
	α	Nd–141m	61 s	·23–2	0·11	s
	2	**Sm–143m**	64 s	·25–2	2·20	sex
	p	**Pm–154**	2·5 m	·62–2	0·54	
	p	**Pm–152**	6·5 m	·03–1	0·80	
	2	**Sm–143°**	8·3 m	·14–1	2·20	sex
	α	Nd–151	12 m	·30–1	0·04	
	γ	Sm–155	24 m	·60–1	·55–1	ex
	α	Nd–149	2 h	0·30	1·03	ex
	α	Nd–144°	2·42 h	0·38	0·11	s
	p	**Pm–150**	2·7 h	0·43	0·43	
	—	Pm–151	26·5 h	1·44	—	
	γ	Sm–153	47 h	1·67		
Sm (62)	2	Sm–153	47 h	1·67	3·12	ex
	p	Pm–149	53·1 h	1·73	0·75	
	—	Pm–149	53·1 h	1·73	0·75	
	p	Pm–148′	5·3 d	2·10	0·74	s
	α	Nd–147	11·3 d	2·44	·91–1	
	p	Pm–148″	42 d	3·00	0·74	s
	—	**Pm–143**	~300 d	3·85	—	
	γ	**Sm–145**	360 d	3·98		
	p	**Pm–144**	~450 d	4·03	0·51	
	—	Eu–155	1·7 y	4·17	—	
	p	Pm–147	2·52 y	4·34	0·93	
	p	Pm–147	2·52 y	4·34	—	
	—	**Pm–145**	18 y	5·20	—	
	γ	Sm–151	~100 y	5·94		
	2	Sm–151	~100 y	5·94	3·15	
	—	Sm–151	~100 y	5·94	—	
	γ	Eu–152′	9·2 h	0·96	6·43	ex
	γ	Eu–152″	13 y	5·05	7·12	ex
	γ	Eu–154	16 y	5·14	5·94	ex
	α	Pm–50	2·7 h	0·43	0·46	
	γ	Eu–152′	9·2 h	0·96		
	2	Eu–152′	9·2 h	0·96	2·53	
Eu (63)	2	**Eu–150**	13·7 h	1·13	3·09	ex
	p	Sm–153	47 h	1·67	1·19	ex
	α	Pm–148′	5·3 d	2·10	·58–1	s
	α	Pm–148″	42 d	3·00	·58–1	s
	γ	Eu–152″	13 y	5·05		
	2	Eu–152″	13 y	5·05	3·40	
	γ	Eu–154	16 y	5·74		
	p	Sm–151	~100 y	5·94	0·11	
	γ	Gd–161	3·63 m	·78–2	2·82	ex
	γ	Gd–159	18 h	1·25	3·58	ex
Gd (64)	—	Tb–161	7 d	2·22	—	
	γ	Gd–153	236 d	3·75	<3·00	ex
	α	**Sm–157**	30 s	·92–3	·91–1	
	p	**Eu–160**	2·5 m	·62–2	0·52	
	γ	Gd–161	3·63 m	·78–2	1·17	ex
	α	Sm–155	24 m	·60–1	0·15	
	p	**Eu–158**	60 m	0·00	0·67	
	p	Eu–152′	9·2 h	0·96	·04–1	s
	p	**Eu–157**	15·4 h	1·18	0·82	ex
	—	**Eu–157**	15·4 h	1·18	—	
	γ	Gd–159	18 h	1·25		
	2	Gd–159	18 h	1·25	3·09	ex
Gd (64)	α	Sm–153	47 h	1·67	0·40	ex
	—	Tb–161	7 d	2·22	—	
	p	Eu–156	15·4 d	2·57	0·74	
	2	**Gd–151**	150 d	3·55	0·92	
	γ	Gd–153	236 d	3·75		
	2	Gd–153	236 d	3·75	1·98	
	p	Eu–155	1·7 y	4·17	0·66	
	—	Eu–155	1·7 y	4·17	—	
	p	Eu–152″	13 y	5·05	·04–1	s
	p	Eu–154	16 y	5·14	·92–1	
	α	Sm–151	~100 y	5·94	·43–1	
	γ	Tb–160	73 d	3·24	5·22	ex
	2	Tb–158m	11 s	·48–3	3·68	s
Tb (65)	p	Gd–159	18 h	1·25	1·18	
	α	Eu–156	15·4 d	2·56	0·65	
	γ	Tb–160	73 d	3·24	2·08	ex
	2	Tb–158°	1 ky	6·94	3·68	s
	γ	Dy–165m	1·25 m	·32–2	6·32	ex
	γ	Dy–165°	2·32 h	0·36	5·92	ex
	γ	**Dy–157**	8·2 h	0·91		
	—	Tb–157	>30 y	5·42	—	
	γ	Dy–159	144 d	3·54	2·52	ex
	p	Tb–158m	11 s	·48–3	·61–2	s
	γ	Dy–165m	1·25 m	·32–2		
	α	Gd–161	3·63 m	·78–2	0·08	
	p	**Tb–163**	7 m	·06–1	0·66	
	p	Tb–162	14 m	·37–1	0·76	
	γ	Dy–165°	2·32 h	0·36		
Dy (66)	p	**Tb–156′**	5·5 h	0·74	·51–2	s
	γ	**Dy–157**	8·2 h	0·91		
	2	**Dy–157**	8·2 h	0·91	0·58	
	2	**Dy–155**	10 h	1·00	0·30	
	α	Gd–159	18 h	1·25	0·53	ex
	p	**Tb–164**	23 h	1·36	0·67	
	p	**Tb–156″**	5·2 d	2·10	·51–2	s
	—	Tb–155	5·6 d	2·13	—	
	p	Tb–161	7 d	2·23	0·69	
	—	Tb–161	7 d	2·23	—	
	p	Tb–160	73 d	3·24	·84–1	
	γ	Dy–159	144 d	3·54		
	2	Dy–159	144 d	3·54	2·00	
	α	Gd–153	236 d	3·75	·04–2	

Target		Product	$t_{\frac{1}{2}}$	log $t_{\frac{1}{2}}$ [h]	log S_∞ [dps/mg]		Target		Product	$t_{\frac{1}{2}}$	log $t_{\frac{1}{2}}$ [h]	log S_∞ [dps/mg]	
Dy (66)	—	Tb–157	>30 y	5·42	—			p	Er–169	9·4 d	2·35	1·15	
	—	Tb–157	>30 y	5·42	—		Tm	2	Tm–168	85 d	3·31	3·66	
	p	Tb–158°	1 ky	6·94	·61–2	s	(69)	γ	Tm–170	127 d	3·48	1·63	ex
								α	Ho–166″	>30 y	5·42	0·63	s
	γ	Ho–166′	27 h	1·43	5·37	ex							
	γ	Ho–166″	>30 y	5·42				γ	Yb–177m	6·5 s	·26–3		
	p	Dy–165m	1·25 m	·32–2	1·04	s		γ	**Yb–169m**	46 s	·11–2		
Ho	α	Tb–162	14 m	·37–1	0·57			γ	Yb–177°	1·9 h	0·27	3·38	ex
(67)	2	Ho–164	37 m	·79–1	3·88	ex		γ	Yb–175	4·2 d	2·00	~4·82	ex
	p	Dy–165°	2·32 h	0·36	1·04	s		—	Lu–177	6·7 d	2·20	—	
	γ	Ho–166′	27 h	1·43				γ	**Yb–169°**	32 d	2·88	4·74	ex
	γ	Ho–166″	>30 y	5·42				α	**Er–173**			·59–1	
								α	Er–167m	2·5 s	·84–4	·43–1	
	γ	Er–167m	2·5 s	·62–4				γ	Yb–177m	6·5 s	·26–3		
	γ	**Er–163**	75 m	0·09	1·01	ex		γ	**Yb–169m**	46 s	·11–2		
	γ	Er–171	7·5 h	0·87	3·68	ex		2	**Yb–169m**	46 s	·11–2	2·11	s
	γ	Er–165	10 h	1·00	1·98	ex		p	**Tm–176**	2 m	·52–2	0·18	
	γ	Er–169	9·4 d	2·35	3·29	ex		p	**Tm–174**	5·5 m	·96–2	0·71	
	—	Tm–171	1·9 y	4·22	—			2	Yb–167	18·5 m	·49–1	0·76	
	—	**Ho–163m**	0·8 s	·35–4	—		Yb	γ	Yb–177°	1·9 h	0·27		
	—	**Ho–163m**	0·8 s	·35–4	—		(70)	p	Tm–173	7·2 h	0·86	0·23	
	γ	Er–167m	2·5 s	·62–4				—	Tm–173	7·2 h	0·86	—	
	2	**Er–167m**	2·5 s	·62–4	3·11			α	Er–171	7·5 h	0·87	0·15	
	—	**Er–167m**	2·5 s	·62–4	—			α	Er–165	10 h	1·00	·28–2	
	—	**Er–167m**	2·5 s	·62–4	—			p	Tm–172	63 h	1·80	0·66	
	p	**Ho–170**	40 s	·05–2	0·32			γ	Yb–175	4·2 d	2·00		
	α	Dy–165m	1·25 m	·32–2	0·18	s		2	Yb–175	4·2 d	2·00	2·54	ex
	p	**Ho–168**	3·3 m	·74–2	0·69			—	Lu–177	6·7 d	2·20	—	
	α	**Dy–167**	4·4 m	·87–2	·72–1			α	Er–169	9·4 d	2·35	0·15	
	p	Ho–164	37 m	·79–1	·76–1			—	**Tm–167**	9·6 d	2·36	—	
Er	p	**Ho–162**	67 m	0·05	·82–2			γ	**Yb–169°**	32 d	2·88		
(68)	γ	**Er–163**	75 m	0·09				2	**Yb–169°**	32 d	2·88	2·11	s
	2	**Er–163**	75 m	0·09	1·83			p	Tm–168	85 d	3·31	·74–2	
	α	Dy–165°	2·32 h	0·36	0·18	s		p	Tm–170	127 d	3·48	·93–1	
	—	**Ho–161**	2·5 h	0·39	—			p	Tm–171	1·9 y	4·22	0·54	
	p	**Ho–167**	3 h	0·48	0·70			—	Tm–171	1·9 y	4·22	—	
	—	**Ho–167**	3 h	0·48	—								
	2	**Er–161**	3·5 h	0·54	0·73			γ	Lu–176m	3·71 h	0·57	5·07	ex
	γ	Er–171	7·5 h	0·87				γ	Lu–177	6·7 d	2·20	5·55	ex
	γ	Er–165	10 h	1·00			Lu	γ	Lu–176m	3·71 h	0·57		
	2	Er–165	10 h	1·00	3·18		(71)	α	Tm–173	7·2 h	0·86		
	p	Ho–166′	27 h	1·43	0·93	s		α	Tm–172	63 h	1·80	0·53	
	γ	Er–169	9·4 d	2·35				p	Yb–175	4·2 d	2·00	1·06	ex
	2	Er–169	9·4 d	2·35	2·89			γ	Lu–177	6·7 d	2·20		
	α	Dy–159	144 d	3·54	·36–2			2	Lu–174	165 d	3·59	3·65	
	—	Tm–171	1·9 y	4·22	—								
	—	**Ho–163°**	>20 y	5·24	—			γ	**Hf–178m**	4·8 s	·12–3		
	—	**Ho–163°**	>20 y	5·24	—		Hf	γ	Hf–179m	19 s	·72–3	4·84	ex
	p	Ho–166″	>30 y	5·42	0·93	s	(72)	γ	Hf–180m	5·5 m	0·74	4·48	ex
Tm	γ	Tm–170	127 d	3·48	5·62	ex		γ	Hf–181	46 d	3·04	4·13	ex
(69)	α	Ho–166′	27 h	1·43	0·63	s							

SOME BASIC PRINCIPLES

Target		Product	$t_{\frac{1}{2}}$	log $t_{\frac{1}{2}}$ [h]	log S_∞ [dps/mg]	
	γ	Hf–175	70 d	3·22	2·99	ex
	γ	**Hf–178m**	4·8 s	·12–3		
	2	**Hf–178m**	4·8 s	·12–3		
	α	Yb–177m	6·5 s	·26–3	0·30	s
	γ	Hf–179m	19 s	·72–3		
	2	Hf–179m	19 s	·72–3	3·23	
	—	Hf–179m	19 s	·72–3	—	
	p	**Lu–180**	4·5 m	·88–2	0·67	
	p	Lu–178	22 m	·56–1	0·66	
	α	Yb–177°	1·9 h	0·27	0·30	s
Hf	p	Lu–176m	3·71 h	0·57	0·08	
(72)	γ	Hf–180m	5·5 h	0·74		
	—	Hf–180m	5·5 h	0·74	—	
	p	**Lu–179**	8 h	0·90	0·36	
	2	**Hf–173**	23·6 h	1·37	0·87	
	α	Yb–175	4·2 d	2·00	0·26	ex
	p	Lu–177	6·7 d	2·20	0·58	
	—	Lu–177	6·7 d	2·20	—	
	γ	Hf–181	46 d	3·04		
	γ	**Hf–175**	70 d	3·22		
	2	**Hf–175**	70 d	3·22	2·36	
	p	Lu–174	165 d	3·59	·79–2	
	—	**Lu–173**	~200 d	3·67	—	
	γ	Ta–182m	16·5 m	·44–2	2·00	ex
	γ	Ta–182°	112 d	3·43	4·80	ex
	γ	Ta–182m	16·5 m	0·44–2	1·52	
	α	Lu–178	22 m	·56–1	0·43	
Ta	p	Hf–180m	5·5 h	0·74		
(73)	2	Ta–180m	8·15 h	0·91	3·48	ex
	α	Lu–177	6·7 d	2·20		
	p	Hf–181	46 d	3·04	1·00	
	γ	Ta–182°	112 d	3·43		
	2	Ta–179	600 d	4·16		
	γ	W–183m	5·5 s	·18–3	5·24	sex
	α	W–185m	1·7 m	·45–2		
	γ	W–187	24 h	1·38	4·49	ex
	γ	W–185°	74 d	3·25	3·34	ex
	γ	W–181	140 d	3·52	~2·13	ex
	γ	W–183m	5·5 s	·18–3		
W	2	W–183m	5·5 s	·18–3	3·15	s
(74)	α	Hf–179m	19 s	·72–3	0·20	
	γ	W–185m	1·7 m	·45–2		
	2	W–185m	1·7 m	·45–2	3·15	s
	2	**W–179m**	6 m	·00–1	0·75	s
	p	**Ta–186**	10·5 m	·24–1	0·55	ex
	p	Ta–182m	16·5 m	·44–1	0·72	s
	2	**W–179°**	40 m	·82–1	0·75	s

Target		Product	$t_{\frac{1}{2}}$	log $t_{\frac{1}{2}}$ [h]	log S_∞ [dps/mg]	
	α	**Hf–183**	64 m	0·03	·92–1	
	α	Hf–180m	5·5 h	0·74		
	p	Ta–180m	8·15 h	0·91	·56–2	
	p	Ta–184	8·7 h	0·94	0·70	ex
	γ	W–187	24 h	1·38	0·57	ex
W	p	**Ta–183**	5·2 d	2·09	0·38	
(74)	α	Hf–181	46 d	3·04	0·11	
	γ	W–185°	74 d	3·25		
	2	W–185°	74 d	3·25	3·15	s
	p	Ta–182°	112 d	3·43	0·72	s
	γ	W–181	140 d	3·52		
	2	W–181	140 d	3·52	3·08	
	—	Ta–179	600 d	4·16	—	
	γ	Re–188m	18·7 m	·49–1		
	γ	Re–188°	17 h	1·22	5·15	ex
	γ	Re–186	3·79 d	1·96	5·16	ex
	p	W–185m	1·7 m	·45–2	0·62	s
	α	Ta–182m	16·5 m	·44–1	0·08	s
	γ	Re–188m	18·7 m	·49–1		
Re	α	Ta–184	8·7 h	0·94	0·26	ex
(75)	γ	Re–188°	17 h	1·22		
	p	W–187	24 h	1·38	0·91	ex
	2	Re–184′	2·2 d	1·72	3·20	s
	γ	Re–186	3·79 d	1·96		
	2	Re–186	3·79 d	1·96	3·70	ex
	2	Re–184″	50 d	3·08	3·20	s
	p	W–185°	74 d	3·25	0·62	s
	α	Ta–182°	112 d	3·43	0·08	s
	—	Ir–191m	4·9 s	·14–3	—	
	γ	Os–190m	10 m	·22–1		
	γ	Os–189m	5·7 h	0·76		
	γ	Os–191m	14 h	1·14		
	γ	Os–193	31·5 h	1·50	3·31	ex
	γ	Os–191°	16 d	2·58	3·83	ex
	γ	Os–185	95 d	3·36	<2·07	ex
	α	**W–189**			0·00	
	—	Ir–191m	4·9 s	·14–3	—	
Os	—	Ir–191m	4·9 s	·14–3	—	
(76)	α	W–183m	5·5 s	·18–3	·00–1	
	α	W–185m	1·7 m	·45–2	·81–1	s
	p	Re–190	2·8 m	·66–2	0·52	
	p	**Re–192**	9·8 m	·21–1	0·59	
	γ	Os–190m	10 m	·22–1		
	p	Re–188m	18·7 m	·50–1	0·32	s
	γ	Os–189m	5·7 h	0·76		
	2	Os–189m	5·7 h	0·76	3·08	
	2	**Os–183′**	10 h	1·00	·85–1	s
	γ	Os–191m	14 h	1·14		

Target		Product	$t_{1/2}$	log $t_{1/2}$ [h]	log S_∞ [dps/mg]	
Os (76)	2	Os-191m	14 h	1.14	3.30	s
	2	**Os-183"**	15 h	1.17	·85-1	s
	p	Re-188°	17 h	1.22	0.32	s
	α	W-187	24 h	1.37	·96-1	
	γ	Os-193	31.2 h	1.50		
	p	Re-184'	2.2 d	1.72	·72-3	s
	p	Re-186	91 h	1.96	·56-1	
	γ	Os-191°	16 d	2.58		
	2	Os-191°	16 d	2.58	3.30	s
	p	Re-184"	50 d	3.08	·72-3	s
	—	**Re-183**	70 d	3.22	—	
	—	**Re-183**	70 d	3.22	—	
	α	W-185°	74 d	3.25	·81-1	s
	γ	**Os-185**	95 d	3.36		
	2	**Os-185**	95 d	3.36	1.82	
	α	W-181	140 d	3.53	·23-3	
	p	**Re-189'**	200 d	3.68	0.42	s
	p	**Re-180"** >1 ky		6.94	0.42	s
Ir (77)	γ	Ir-192m	1.45 m	·38-2	5.50	ex
	γ	Ir-194	19 h	1.27	5.40	ex
	γ	Ir-192°	74.5 d	3.25	5.93	ex
	—	Ir-191m	4.9 s	·14-3	—	
	γ	Ir-192m	1.45 m	·38-2		
	2	Ir-192m	1.45 m	·38-2	3.46	s
	α	Re-190	2.8 m	·67-2	0.08	
	—	Os-190m	10 m	·22-1	—	
	—	Os-190m	10 m	·22-1		
	α	Re-188m	18.7 m	·49-1	0.48	ex
	2	Ir-190'	3.2 h	0.50	2.65	ex
	p	Os-191m	14 h	1.14	0.56	s
	α	Re-188°	17 h	1.22		
	γ	Ir-194	19 h	1.27		
	p	Os-193	31.5 h	1.49	0.71	ex
	2	Ir-190"	12 d	2.46		
	p	Os-191°	16 d	2.58	0.56	s
	γ	Ir-192°	74.5 d	3.25		
	2	Ir-192°	74.5 d	3.25	3.46	s
Pt (78)	γ	Pt-199m	14 s	·59-3		
	γ	Pt-199°	31 m	·71-1	2.94	ex
	γ	Pt-197m	80 m	0.12		
	γ	Pt-197°	18 h	1.25	2.83	ex
	γ	Pt-191	3 d	1.86	1.76	ex
	—	Au-199	3.15 d	1.88	—	
	γ	Pt-193m	3.5 d	1.92	2.29	sex
	γ	Pt-195m	6 d	2.16	3.11	ex
	γ	Pt-193°	<500 y	6.64	2.29	sex
	—	Ir-191m	4.9 s	·14-3	—	
	γ	Pt-199m	14 s	·59-3	·57-1	sex
	p	**Ir-198**	50 s	·14-2	·18-1	
	p	Ir-192m	1.45 m	·38-2	·18-1	s
	α	**Os-195**	6.5 m	·03-1	·82-1	
	—	Os-190m	10 m	·22-1	—	
	γ	Pt-199°	31 m	·71-1	·57-1	sex
	γ	Pt-197m	80 m	0.12		
	2	Pt-197m	80m	0.12		
	p	**Ir-195**	2.3 h	0.36	0.48	ex
	—	**Ir-195**	2.3 h	0.36	—	
	p	Ir-190'	3.2 h	0.51	·48-3	s
	p	**Ir-196**	<5 h	0.70	0.49	
	α	Os-189m	5.7 h	0.76	·64-2	
	—	Os-189m	5.7 h	0.76	—	
	2	**Pt-189**	10.5 h	1.02	·67-1	
Pt (78)	α	Os-191m	14 h	1.14	0.12	sex
	γ	Pt-197°	18 h	1.25		
	2	Pt-197°	18 h	1.25	2.79	ex
	p	Ir-194	19 h	1.27	0.60	ex
	α	Os-193	31.5 h	1.50	·63-1	ex
	γ	**Pt-191**	3 d	1.86		
	2	**Pt-191**	3 d	1.86	1.52	
	—	Au-199	3.15 d	1.88	—	
	γ	Pt-193m	3.5 d	1.92		
	2	Pt-193m	3.5 d	1.92	3.18	s
	γ	Pt-195m	6 d	2.16		
	2	Pt-195m	6 d	2.16	3.08	
	—	**Ir-189**	11 d	2.42	—	
	p	Ir-190"	12 d	2.46	·48-3	s
	α	Os-191°	16 d	2.58	0.12	sex
	p	Ir-192°	74.5 d	3.25	·18-1	s
	γ	Pt-193°	<500 y	6.64		
	2	Pt-193°	<500 y	6.64	3.18	s
Au (79)	γ	Au-198	2.7 d	1.81	5.48	ex
	p	Pt-197m	80 m	0.12	0.87	sex
	2	Au-196m	10 h	1.00	3.90	sex
	p	Pt-197°	18 h	1.25	0.87	sex
	α	Ir-194	19 h	1.27	0.18	ex
	γ	Au-198	2.7 d	1.81	~1.37	
	2	Au-196°	5.6 d	2.13	3.74	ex
	—	**Au-197m**	7.4 s	·31-3	—	
Hg (80)	γ	Hg-205	5.5 m	·96-2	1.94	ex
	γ	Hg-199m	44 m	·87-1	0.74	ex
	γ	**Hg-197m**	24 h	1.37	3.27	ex
	γ	**Hg-197°**	65 h	1.81	3.60	ex
	γ	Hg-203	45.8 d	3.04	3.53	ex
	p	**Au-204**			·74-1	
	—	**Au-197m**	7.4 s	·31-3	—	
	—	**Au-197m**	7.4 s	·31-3	—	
	α	Pt-199m	14 s	·59-3	·95-1	sex

SOME BASIC PRINCIPLES

Target		Product	$t_{\frac{1}{2}}$	log $t_{\frac{1}{2}}$ [h]	log S_∞ [dps/mg]		Target		Product	$t_{\frac{1}{2}}$	log $t_{\frac{1}{2}}$ [h]	log S_∞ [dps/mg]	
	p	Au–202	25 s	·84–3	0·43			γ	Tl–206	4·2 m	·84–2	0·62	ex
	—	Au–195m	30 s	·92–3	—			p	Hg–205	5·5 m	·96–2	0·79	ex
	α	Pt–201	2·3 m	·52–2	·08–1		Tl	α	Au–200	48 m	·90–1	·54–1	ex
	γ	Hg–205	5·5 m	·96–2			(81)	2	Tl–202	12 d	2·45	3·11	
	p	Au–201	26 m	·63–1	·92–1	ex		p	Hg–203	45·8 d	3·04	0·34	
	—	Au–201	26 m	·63–1	—			γ	Tl–204	4·1 y	4·56		
	α	Pt–199°	31 m	·71–1	·95–1	sex		2	Tl–204	4·1 y	4·56	3·52	
	γ	Hg–199m	44 m	·87–1				γ	Pb–207m	·84 s	·37–4	1·24	ex
	2	Hg–199m	44 m	·87–1	3·00			γ	Pb–209	3·32 h	0·52	·96–1	ex
	p	Au–200	48 m	·90–1	0·40	ex		γ	Pb–205	30 My	11·42	1·55	ex
	α	Pt–197m	80 m	0·12	0·10	sex		γ	Pb–207m	·84 s	·37–4		
	2	Hg–195°	9·5 h	0·98	0·76	s		2	Pb–207m	·84 s	·37–4	3·40	s
	p	Au–196m	10 h	1·00	·49–2	s		2	Pb–203m	6·7 s	·27–3	1·80	s
Hg	α	Pt–197°	18 h	1·25	0·10	sex		p	Tl–208	3·1 m	·71–2	0·18	ex
(80)	γ	Hg–197m	24 h	1·32			Pb	p	Tl–206	4·2 m	·84–2	0·44	
	2	Hg–197m	24 h	1·37	2·62	s	(82)	p	Tl–207	4·78 m	·90–2	0·30	
	2	Hg–195m	40 h	1·60	0·76	s		α	Hg–205	5·5 m	·96–2	0·38	ex
	p	Au–198	2·7 d	1·81	0·23			γ	Pb–209	3·32 h	0·52	0·66	ex
	γ	Hg–197°	65 h	1·81				2	Pb–203°	52 h	1·72	1·80	s
	2	Hg–197°	65 h	1·81	2·62	s		α	Hg–203	45·8 d	3·04	·84–1	
	p	Au–199	3·15 d	1·88	0 40			p	Tl–204	4·1 y	4·56	·34–]	
	—	Au–199	3·15 d	1·88				γ	Pb–205	30 My	11·42		
	α	Pt–193m	3·5 d	1·92	·97–3	s		2	Pb–205	30 My	11·42	3·00	
	p	Au–196°	5·6 d	2·13	·49–2	s		γ	Bi–210′	5 d	2·03	1·64	ex
	α	Pt–195m~	6 d	2·16	·66–1			γ	Bi–210″	2·6 My	10·36	1·74	ex
	γ	Hg–203	45·8 d	3·04			Bi	α	Tl–206	4·2 m	·84–2	0·16	ex
	2	Hg–203	45·8 d	3·04	2·53		(83)	p	Pb–209	3·32 h	0·52	0·57	ex
	—	Au–195°	185 d	3·64	—			γ	Bi–210′	5 d	2·08	0·64	ex
	α	Pt–193°	<500 y	6·64	·97–3	s		2	Bi–208	20 ky	8·24	3·84	ex
Tl	γ	Tl–206	4·2 m	·84–2	2·32	ex		γ	Bi–210″	2·6 My	10·36		
(81)	γ	Tl–204	4·1 y	4·56	3·98	ex							
	α	Au–202	25 s	·84–3	0·00								

^{28}Al is in fact the most abundant activity produced by the bombardment of silicon or of phosphorus with 14 MeV neutrons and would therefore be useful in the estimation of silicon or phosphorus. The presence of ^{28}Al in an irradiated sample cannot be attributed to the (n, γ) reaction on ^{27}Al unless it is known that phosphorus and silicon are absent from the sample.

Interfering reactions can often be eliminated by irradiating the sample with neutrons at energies below the threshold level for the relevant reaction. Compilations of cross-sections for various nuclear reactions as a function of neutron energy are generally available in the literature. (Howerton, 1958; Hughes and Schwartz, 1958). The exploitation of threshold levels is however limited by the difficulty of producing adequate neutron fluxes in the energy ranges desired. Some reactions useful in this regard are noted in Table IV. Neutrons of

TABLE IV

Nuclear Reactions used to Produce Neutrons of Defined Energy (after Lyon, 1964)

Nuclear reaction	Neutron energy
$^{12}C(d, n)^{13}N$	3·4 keV
$^{2}H(d, n)^{3}He$	2·45 MeV
$^{9}Be(\alpha, n)^{12}C$	5·27 MeV
$^{3}H(d, n)^{4}He$	14·05 MeV

various energies can sometimes be produced from the same accelerator, by suitable change of target (Steele, 1965).

6. ACTIVATION WITH CHARGED PARTICLES

For a number of elements, neutron activation analysis—at whatever energy—is not sufficiently sensitive to compete with other techniques. This limitation applies to the lightest nuclei, up to $Z = 8$, and for some heavier nuclei including iron and lead.

In these circumstances it is worth while to consider the possibility of activation analysis with charged particles. Such projectiles have a very limited range in solid matter and it is, therefore, not possible to achieve uniform irradiation unless an extremely thin sample is used. A further difficulty arises because the energy spectrum of a beam of charged particles changes with depth of penetration into the analytical sample. It is not possible, therefore, to make calculations on the basis of a constant cross-section, as can normally be done for neutron activation. Engelmann (1964) has developed techniques to overcome these obstacles, relating measurements on samples and on standards for comparison by a method of equivalent thicknesses.

Though protons, deuterons and α-particles are the most commonly used, ^{3}He particles have attractive properties, first investigated by Markowitz and Mahony (1962). ^{3}He nuclei are transformed rather easily into ^{4}He by neutron capture. For this reason, projectiles of modest energy are suitable; useful activation analysis is possible in many nuclei (up to $Z = 20$) with ^{3}He particles of energy 8 MeV. Bombardment of a target nucleus with ^{3}He may be followed by emission of a neutron, a proton, an α-particle or a photon; (^{3}He, pn) and (^{3}He, αn) reactions are also known. Most of these processes result in the production of neutron-deficient nuclides which are of course positron emitters,

Ricci and Hahn (1965) report limits of detection lying between one part per million and one part per billion for Be, C, N, O and F.

An incidental result of these analytical studies was the finding that an intense neutron flux may be generated by bombardment of thick targets of lithium hydride, beryllium or carbon. Using a $100\,\mu A$ beam of ^{3}He ions at energy 20 MeV (as can be obtained from a 30 in. cyclotron) neutron fluxes between 10^{12} and 10^{13} $n\,\mathrm{cm}^{-2}\,\mathrm{sec}^{-1}$ have been obtained. With the increasing availability of small cyclotrons, the analytical usefulness of ^{3}He ions and other charged particles should be further developed before long.

7. ACTIVATION BY PHOTONS

The average binding energy of a nucleon (neutron or photon) is about 8 MeV; expulsion of a nucleon may therefore be expected if that amount of energy is supplied to the nucleus. When particles are used for the bombardment, rearrangement of the reacting masses often contributes to the release of the required energy, but photon activation is also feasible.

High energy γ-ray quanta are most easily obtained in the form of bremsstrahlung from targets of high atomic number, bombarded by accelerated electrons. The reactions most easily induced by photons are of the form (γ, γ'), (γ, p) and (γ, n). (γ, γ') reactions are readily induced by photons of energy as low as 3 MeV (Lukens et al., 1961), leaving the target nucleus in an isomeric state which undergoes radioactive decay. Such reactions are generally not suitable for activation analysis with reasonable sensitivity. (γ, n) and (γ, p) reactions are more serviceable in this connection, but require photon energies of above 20 MeV. Though in favourable circumstances, excellent sensitivity can be achieved, (Engelmann, 1964, was able to measure oxygen at concentrations of one part per million), the target generally supports a variety of nuclear reactions and careful choice of photon energy is necessary to reduce interference.

C. Assay of Induced Radioactivity

The design of an activation analysis experiment is materially influenced by the equipment available for the ultimate assay of induced activity. Before the development of reliable scintillation techniques, only the Geiger counter was available. Simultaneous determination of two β-emitters was seldom possible; identification and estimation of multiple activities could be achieved only by absorption measurements or by the analysis of decay curves.

When counting techniques were severely limited, radiochemical separation methods were attractive. By application of the traditional analytical techniques including carrier separations to the irradiated sample, it was found possible to determine trace constituents without the delicate and tedious manipulations commonly associated with microchemical analysis and with complete freedom from the reagent contamination potentially hazardous in almost all other techniques. Even now, post-irradiation chemical separation provides the highest sensitivity, with the additional advantage that systematic treatment in this way allows a virtually complete survey of the trace element content of the sample.

A common procedure involves the simultaneous irradiation of the sample under investigation and a standard of known content. The standard must be treated in exactly the same way as the sample in regard to irradiation, subsequent chemical treatment and radioactive assay. The masses of the separated trace elements are then proportionate to the corresponding count rates in the sample and in the standard. This technique is usually favoured when the best possible sensitivity is required. The most satisfactory results are achieved for elements such as arsenic and mercury in which long-lived activities are induced by neutron bombardment, since ample time is then available for post-irradiation chemical treatment. When shortlived activities are to be assayed, the choice of analytical techniques in the radiochemical separation process may be severely limited; useful work has nevertheless been done on activities with half-lives down to a few seconds.

D. Problems of Standardization

The use of a standard sample implies that very small amounts (usually well below 1 mg) appropriate for trace determination can be prepared with adequate accuracy. Microgram quantities of the standard element or compound can be manipulated without danger of loss only if diluted in an inert matrix substance of considerably greater bulk. It is necessary, therefore, to ensure that the matrix contains no trace of the element under investigation; the necessary purity can usually be checked by a separate activation analysis. Homogeneity of distribution of the standard in the supporting matrix is also important, particularly if only an aliquot of a larger preparation is irradiated.

In the internal standard method of Leliaert *et al.* (1958) a known small amount of the element to be determined is homogeneously mixed with part of the sample under investigation; another part of the sample is irradiated without this addition. These irradiations need not be simultaneous, if, as is often the case, a second element present in the sample can be used as a flux monitor.

It would, of course, be useful to have an absolute method of activation analysis, obviating the necessity for standards. It is, however, hardly ever possible to obtain sufficiently accurate values for the activating flux and the cross-sections of the relevant nuclear reactions.

E. γ-Ray Spectrometry

Analysis of γ-ray spectra, derived from scintillation detectors, gives the possibility of simultaneous determination of several elements without chemical separation. With the familiar sodium iodide detector, the interpretation of a γ-ray spectrum is often difficult. The spectral contribution from mono-energetic γ-radiation is not, as might be hoped, a single sharp line, but takes the form of a continuum of varying amplitude with a broad peak providing the only means of energy estimation. When, as is often the case, the irradiated sample provides a variety of γ-emission, the spectrum is very complicated and complete identification of all the elements contributing to it, even in a qualitative fashion, is not possible. For these reasons the simultaneous determination of several elements in an irradiated sample by γ-ray spectroscopy is subject to severe restrictions. In principle, not more than 30 peaks can be distinguished in the range from 100 KeV to 2 MeV; it is in this range that most of the γ-ray quanta emitted by radionuclides are found. In practice, the resolution available is even less. It is often possible to improve the efficiency of identification by measuring the half-lives of appropriate photo-peaks in the γ-ray spectrum. A further improvement in overall resolution is achieved if partial chemical separation is performed before the γ-ray spectrum is plotted. It is, however, virtually impossible to obtain a complete survey of the composition of a sample by activation analysis and γ-ray spectroscopy. For the estimation of a single trace element (or a small number of elements whose γ-ray spectra do not overlap to a confusing extent), non-destructive activation analysis can often be achieved with adequate sensitivity for process control or other routine analyses.

F. Coincidence Methods

Coincidence techniques, depending on recognition of the simultaneity of two or more nuclear events have been widely used for many years in nuclear physics and have been particularly important in the study of elementary particles. Techniques of this kind can usefully improve the energy resolution available in γ-ray spectrometry.

Most of our knowledge of decay schemes of radioactive nuclei is based on the study of $(\beta\gamma)$ or $(\gamma\gamma)$ coincidence measurements during the last quarter century; it is not surprising that analysts have used the same experimental approach to improve the selectivity of trace element estimations without the necessity of chemical manipulation. It is not surprising that the limits of detection by coincidence techniques may be even better than a straightforward γ-ray spectrometry; this improvement is possible because the uncertainty associated with background effects is much smaller in coincidence spectroscopy.

Compilations of analytically useful coincidences, which are evident in the decay schemes of many radionuclides, have been made by Wahlgren et al. (1965) for thermal and by Schulze (1965) for fast neutron activation analysis.

The semi-conductor detectors which have recently become available (Girardi and Guzzi, this volume pp. 137–161) offer very substantial improvement in resolution, though at the expense of sensitivity. It is, however, to be expected that continuing improvements in technology will lead to the production of solid state detectors with counting efficiencies comparable with those offered by the sodium iodide detectors.

G. Limits of Detection

The limit of detection for a given element by γ-ray spectrometry is influenced by a number of factors. If a complete chemical separation has been performed, the γ-ray spectrum will usually be plotted only to verify the identification of the element or the integrity of the chemical manipulation. The total γ-count rates (from sample and standard) will then be used for the numerical estimation of the element under study.

The background count rate, produced by cosmic radiation and inherent radioactive contamination of the background and the experimental equipment, has a significant effect on the attainable limit of analytical detection. When the background is high, or the sample counting rate low, long counting times would be necessary. As a working rule, the net counting rate (after subtraction of background) should be at least three times the standard deviation of the background if significant results are to be obtained.

Numerical estimation of a particular element contributing to a γ-ray spectrum is commonly based on measurement of the area under the corresponding photo peak. The counting channels of pulse height analyser which correspond to the photo peak will not however be filled exclusively by radiations from the γ-ray line on which attention is concentrated. The photo peak is augmented in a complicated way by counts contributed by other radionuclides in the irradiated sample and by other γ-radiation from the nuclide under investigation.

A reasonable estimate of the limit of detection for a particular element can usually be made if the composition of the remainder of the analytical sample is roughly known and if appropriate details of the crystal and counting equipment are specified.

H. Comparison with other Methods of Trace Analysis

It is not easy to make an effective comparison of activation analysis with other techniques. In general, however, spectrophotometry and mass spectrometry are more attractive than activation analysis to the extent that they permit a very rapid survey of a great many elements in a sample. Surface contamination cannot however be excluded in either of these techniques with the same confidence as in activation analysis. A sample subjected to activation analysis may well be contaminated on its surface. If, after the end of the irradiation, the surface is cleaned by acids or organic solvents, surface contamination can be completely removed with no possibility of confusion through contamination by reagents. The reagents may indeed introduce fresh contamination—but not in the form of the radioactive nuclides which are the basis of the subsequent activation analysis. Such freedom from reagent blanks is not easily achieved in spectrophotometry or mass spectrometry.

Though masses down to 10^{-16}g can be detected by mass spectrometry (Specker, 1966), while activation analysis seldom reaches a sensitivity of 10^{-14}g, it is usually not difficult to increase the size of the sample in activation analysis to bring the amount of the element in question above the limit of sensitivity; this can seldom be done with other analytical techniques.

References

Boyd, G. E. (1949). *Analyt. Chem.* **21**, 335–347.
Crouthamel, C. E. (1960). "Applied Gamma-Ray Spectrometry", p. 71. Pergamon Press, Oxford.
Engelmann, C. (1964). CEA–R 2559.
Guinn, V. P. (1962). *In* "Production and Use of Short-Lived Radioisotopes from Reactors", Vol. II, pp. 3–28. *IAEA*, Vienna.
Howerton, R. J. (1958). UCRL–5226.
Hughes, D. J., and Schwartz, R. B. (1958–60). *BNL*–325 and Suppl. 1.

Leliaert, G., Hoste, J. and Eeckhaut, Z. (1958). *Analytica chim. Acta* **19**, 100–107.
Lukens, H. R., Otvos, J. W. and Wagner, C. D. (1961). *Int. J. appl. Radiat. Isotopes* **11**, 30–37.
Lyon, W. S. Jr. (1964). "Guide to Activation Analysis", p. 34. Van Nostrand, Princeton.
Markowitz, S. S. and Mahony, J. D. (1962). *Analyt. Chem.* **34**, 329–335.
Neeb, K. H., Stöckert, H., and Gebauhr, W. (1966). *Z. analyt. Chem.* **219**, 69–76.
Perkins, R. W. and Robertson, D. E. (1965). *In* "Modern Trends in Activation Analysis", pp. 48–57. College Station, Texas.
Ricci, E. and Hahn, R. L. (1965). *Analyt. Chem.* **37**, 742.
Rubinson, W. (1949). *J. chem. Phys.* **17**, 542–547.
Schulze, W. (1964). *In* "L'Analyse par Radioactivation et ses Applications aux Sciences Biologiques", pp. 85–118. Presses Universitaires de France, Paris.
Schulze, W. (1965). *In* "Modern Trends in Activation Analysis", pp. 272–278. College Station, Texas.
Schulze, W. (1966a). *Z. analyt. Chem.* **221**, 85–100.
Schulze, W. (1966b). *Z. analyt. Chem.* **223**, 1–10.
Specker, H. (1966). *Z. analyt. Chem.* **221**, 33–43.
Steele, E. L. (1965). *In* "Modern Trends in Activation Analysis", pp. 102–106. College Station, Texas.
Wahlgren, M., Wing, J. and Hines, J. (1965). *In* "Modern Trends in Activation Analysis", pp. 134–139. College Station, Texas.

REACTORS AS NEUTRON SOURCES

V. P. GUINN

Technical Director, Activation Analysis Program,
Gulf General Atomic Inc.
San Diego, California, U.S.A.

A. The Thermal-Neutron ^{235}U Fission Reaction	37
B. The Fission Spectrum of Neutron Energies	39
C. Moderators, Reflectors and Coolants	39
D. Critical Mass, Reactor Control and Fuel Elements	45
E. Reactor Neutron Fluxes and Flux-Monitoring Techniques . . .	49
F. Irradiation of Samples in Research Reactors in Neutron Activation Analysis Studies	54
1. Thermal-neutron activations	54
2. Fast-neutron activations	60
3. Provisions for irradiation of samples	62
4. Use of recoil protons	68
5. Use of delayed neutrons	70
6. Analysis via prompt γ-rays	70
G. Use of High-Intensity Reactor Pulses	72
H. New Techniques and New Areas of Application	76
References	78
Bibliography	78

The research-type nuclear reactor is the most prolific known source of thermal neutrons, and hence is of considerable interest and use in the field of neutron activation analysis (NAA). To date, the great majority of published papers in the field of activation analysis have involved results obtained with the high thermal-neutron fluxes (10^{11}–10^{14} n cm^{-2} sec^{-1}) of research-type nuclear reactors. It is the purpose of this chapter to briefly discuss the main features of the modern research-type nuclear reactor—from the standpoint of the activation analyst.

A. The Thermal-Neutron ^{235}U Fission Reaction

Virtually all research reactors operate on the basis of the thermal-neutron induced nuclear fission chain reaction of uranium-235 (although

^{233}U and ^{239}Pu are also suitable fuels). Very shortly after the discovery of the fission reaction in 1939, by O. Hahn and F. Strassman, it was established experimentally that the reaction, in each fission event, not only forms two intermediate mass-number radionuclides, but also typically releases two or three neutrons (for ^{235}U, the average number is 2·42) and a large amount of energy. The total energy release in a fission event, corresponding, from the Einstein mass-energy relationship (E = mc^2), to the mass decrease that occurs (typically about 0·215 atomic mass unit), is mostly emitted promptly in the form of kinetic energy (K.E.) of the neutrons emitted, of the two recoiling fission-product nuclei, and in the prompt γ-radiation emitted. Some of the energy is released over a period of time after the event—by the radioactive decay of the various fission-product radionuclides.

The ^{235}U fission reaction may be written as:

$$^{235}U + n \rightarrow [^{236}U]^* \rightarrow 2\ F.P. + 2\text{-}3\ n + \gamma. \tag{1}$$

In this representation, the [^{236}U]* term denotes the excited compound nucleus immediately resulting from the exoergic capture of a neutron—before fission has occurred. The lifetime of the compound nucleus is extremely short, and it almost immediately disintegrates into two large fragments—the two primary fission-product nuclei, of intermediate mass numbers—and the released neutrons.

A mass decrease of 0·215 a.m.u. corresponds to an energy release of about 200 MeV. This energy is distributed approximately as follows:

Prompt:	K.E. of fission fragments	165 MeV	
	Prompt γ-radiation	8	
	K.E. of fission neutrons	5	
	Sub total		178 MeV
Delayed:	Beta particles from decay of fission products	7	
	γ-Rays from decay of fission products	6	
	Neutrinos from decay of fission products	10	
	Sub total		23
Total			201

The fission-product nuclides consist of a great variety of radionuclides, ranging in mass numbers from about 72 to 117, and from about 117 to 161, i.e. in two groups: the light-weight fragment group and the heavy-weight fragment group. In general, each fission event produces one light-weight radionuclide and one heavy-weight radionuclide, that is, the fissioning is asymmetric, rather than symmetric. The initial light-weight nuclei range from $Z = 30$ (Zn) to $Z = 49$ (In), and the initial heavy-weight nuclei range from $Z = 50$ (Sn) to $Z = 65$ (Tb). The fission-product yield curve is symmetrical about mass number 117 (roughly 234/2), with broad maxima at approximately $A = 95$ and $A = 138$. The primary fission products are very short-lived neutron-rich beta (β^-) emitters. In anywhere from one to seven consecutive β^- decays, these eventually decay to a stable nuclide (Ge to Sn in the lighter group; Sb to Dy in the heavier group).

Some 99·35% of the neutrons formed by the fission of ^{235}U are released promptly (within $\sim 10^{-14}$ sec). However, the remaining 0·65% are released over a period of time of seconds to minutes after the fission event has occurred. These delayed neutrons are of considerable importance in reactor control, and are produced largely from six fission-product nuclei that decay by neutron emission. These have half lives of about 0·05, 0·43, 1·52, 4·51, 22·0, and 55·6 secs.

The prompt γ-radiation emitted by the fission event consists of γ-ray photons ranging in energy from almost 0 up to about 7 MeV, but the average energy is about 1 MeV per photon.

B. The Fission Spectrum of Neutron Energies

The initial energies of the neutrons released by the fission of ^{235}U range from almost 0 up to perhaps 25 MeV. However, the most probable energy is about 0·72 MeV, and the average energy is about 2·0 MeV. The initial fission-spectrum of neutron energies can be represented by the equation (Watt, 1952);

$$N(E) = 0{\cdot}48394\, e^{-E} \sinh (\sqrt{2E}). \tag{2}$$

This is normalized so that $\Sigma N = 1$, when summed over all neutron energies. Above about 3 MeV, a plot of log N(E) versus E is almost linear (the slight curvature being due to the log sinh ($\sqrt{2E}$) term in the logarithmic form of the Watt equation:

$$\log_{10} N(E) = -0{\cdot}316 - 0{\cdot}4343\, E + \log_{10} \sinh (\sqrt{2E}). \tag{3}$$

Very approximately, above about 3 MeV, the flux decreases by a factor of 10 for each 3 MeV increase in neutron energy. A plot of $\log_{10} N(E)$ versus E (E in MeV) is shown in Fig. 1.

C. Moderators, Reflectors and Coolants

The fission cross section of ^{235}U is very large at low neutron energies, and decreases considerably at higher neutron energies. Thus, at thermal energies (~17°C), where the average kinetic energy is 0·025 eV (average of a Maxwellian distribution), σ_f for ^{235}U is 582 barns (1 barn = 10^{-24} cm²/nucleus). At 1 eV, it has declined to only 60 barns. Except for a number of strong resonances, in the energy region from 1 eV to about

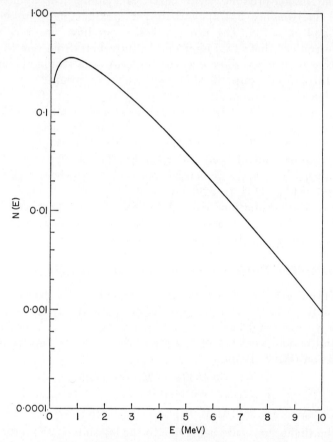

FIG. 1. The fission spectrum of neutron energies.

100 eV, σ_f continue to decline, reaching values of 4·0 barns, 1·8 barns, and 1·2 barns at neutron energies of 0·01, 0·1, and 1 MeV, respectively.

In order to fission ^{235}U with high efficiency, it is thus seen to be desirable to slow down the fission-spectrum neutrons ($\overline{E}$ = 2·0 MeV) to thermal velocities, where σ_f is much greater. This is accomplished by the use of a moderator.

The ideal moderator is a material consisting of nuclei with a mass the same as that of a neutron (to provide the maximum energy loss of a neutron, when colliding with it), and a zero cross section for the absorption (capture) of neutrons of any energy. Although no actual element meets these ideal criteria, some of the lighter (low Z) elements provide reasonable compromises, and are used. In particular, the elements, hydrogen, beryllium, and carbon are used considerably as neutron moderators and reflectors.

The maximum fractional loss of energy that a fast neutron can experience, upon collision with a nucleus of mass number A, is given by the equation,

$$\left(\frac{E_1-E_2}{E_1}\right)_{max} = 1-\left(\frac{A-1}{A+1}\right)^2. \qquad (4)$$

Thus, in this respect, ^{1}H is ideal, since it is possible in a head-on collision for a neutron to give up all of its kinetic energy to a proton. For ^{9}Be and ^{12}C, the maximum possible fractional energy losses by a neutron are 0·360 and 0·284, respectively. Deuterium (^{2}H) is also very good, providing a maximum fractional energy loss of 0·889. Collisions can range all the way from head-on collisions ($\theta = 180°$) to bare glancing collisions ($\theta \sim 0°$), and hence the fractional energy losses in individual events can range all the way from almost zero up to the maximum possible. Integrating over all scattering angles, one can derive a quantity, ξ, the average logarithmic energy decrement per collision, i.e. the average value of $\ln(E_1/E_2)$. The values of ξ for ^{1}H, ^{2}H, ^{9}Be, and ^{12}C are, respectively, 1·000, 0·725, 0·205, and 0·158. From these ξ values, one can compute that the average number of collisions with these moderator nuclei needed to slow down an average fission-spectrum neutron (E = 2·0 MeV) to thermal energy (0·025 eV) equals 18, 25, 86, and 114, respectively. By contrast, for ^{238}U, ξ is 0·00838, and 2172 collisions are required. Helium-4 is also a good moderator, but the fact that it is a gas at all normal temperatures and pressures, and hence has a low density, restricts its practical usefulness as a moderator.

When used as moderators, Be and C are normally used in high-purity elemental form (or Be as BeO), but ^{1}H and ^{2}H are usually employed in the form of high-purity water (H_2O and D_2O). The product of ξ and Σ_s, the macroscopic scattering cross section for epithermal neutrons, is called the slowing down power of the material. The values of the slowing down power of H_2O, D_2O, Be, and C are 1·53, 0·177, 0·16, and 0·063 cm^{-1}, respectively.

Even though a particular material rapidly slows down fast neutrons, it may still not be a useful moderator, i.e. if it absorbs many of the neutrons in the slowing-down process or after they have been thermalized.

For this reason, two other low-Z elements, lithium and boron, are not useful. Lithium has a high atomic thermal-neutron capture cross section, 70·4 barns (ordinary Li is 7·42 atomic % ^{6}Li and 92·58 atomic % ^{7}Li; these isotopes have capture cross sections of 950 barns, and 0·036 barn, respectively). Thus, pure ^{7}Li would be quite useful as a moderator, but the ^{6}Li present in natural lithium makes the cross section much too large for such use. The ^{6}Li exhibits this large cross section for the ^{6}Li(n, α)^{3}H reaction. The situation with boron is quite analogous. Its atomic thermal-neutron capture cross section is 762 barns, due almost entirely to the ^{10}B nuclide, which undergoes the ^{10}B(n, α)^{7}Li reaction with a cross section of 4017 barns. The ^{11}B capture cross section is only 0·005 barn. Thus, pure ^{11}B would be a useful moderator, but not ordinary boron, since the latter consists of 19·8 atomic % ^{10}B and 80·2 atomic % ^{11}B.

The ratio of the slowing down power to the macroscopic absorption cross section, Σ_a, is called the moderating ratio of a material. Thus, it is equal to $\Sigma_s \xi / \Sigma_a$. The moderating ratio values for H_2O, D_2O, Be, and C are, respectively, 70, 21000, 150, and 170. The excellent ratio for D_2O is due primarily to the fact that ^{2}H has an extremely low thermal-neutron capture cross section (0·00057 barn), compared with that of ^{1}H(0·33 barn); the corresponding values for elemental Be and C are 0·010 and 0·0037 barn, respectively.

Beryllium and carbon have the advantage that they are solids, and hence can be machined into desired shapes. Also, they will withstand high temperatures, although both will oxidize at high temperatures, in the presence of O_2: Be forming a protective oxide coating of BeO, C oxidizing to gaseous CO and CO_2. Ordinary water and heavy water have a low boiling point, but, being liquids at normal temperatures, they can be used to fill containers of any desired shapes. The cost of even the distilled, de-ionized water that is used in pool reactors is essentially negligible, whereas the cost of D_2O is appreciable (at present, D_2O costs about one dollar per five grams, in moderate quantities). Thus, in reactors that use D_2O, it is maintained in a closed system—to eliminate loss by vaporization and dilution by atmospheric H_2O vapor. Some provision must then be made to vent the D_2 and O_2 formed by radiolysis, or to recombine them non-explosively.

A parameter that takes into account the amount of space required to slow a fast neutron down to thermal velocity, in a non-absorbing medium, is called the Fermi age. Its values, for the slowing down of an average fission-spectrum neutron (2·0 MeV) to thermal energy (0·025 eV), are 33, 120, 98, and 350 cm^2, respectively, for H_2O, D_2O, Be, and graphite (C). It is related to the mean square distance traveled while slowing down. Thus, the square root of the Fermi age is called the slowing down length (with values, thus, of 5·74, 10·9, 9·90, and 18·7 cm for H_2O, D_2O, Be, and graphite, respectively).

After slowing down to thermal velocity in an infinite medium, a thermalized neutron will diffuse until it is captured. The diffusion length, L, of such thermal neutrons is a parameter of a particular medium. For H_2O, D_2O, Be, and graphite, L has values, respectively, of 2·88, 100, 23·6, and 50·2 cm. The migration length, M, takes into account both the slowing down length and the diffusion length, being equal to the square root of the sum of the squares of these two quantities. The values of M for H_2O, D_2O, Be, and graphite are, respectively, 6·43, 101, 25·8, and 53·6 cm.

The actual length of time required, on the average, for a fast neutron to slow down to thermal velocity—the slowing down time, t—is also a characteristic of the medium, and the initial neutron energy. For the thermalization of 2·0 MeV neutrons to 0·025 eV, the values of t are 1×10^{-5}, $2 \cdot 9 \times 10^{-5}$, $7 \cdot 8 \times 10^{-5}$, and $1 \cdot 9 \times 10^{-4}$ sec, respectively, for H_2O, D_2O, Be, and graphite. Thus, typical thermalization times for an average fission-spectrum neutron range from 10 to 190 μsec, depending upon the material, for these four often-used moderator-reflector materials. Similarly, the length of time that a thermal neutron can diffuse, in an infinite medium, on the average, before it is captured— called the diffusion time, l_0—is a characteristic of the medium. For the same four materials, the values for l_0 are $2 \cdot 1 \times 10^{-4}$, 0·15, $4 \cdot 3 \times 10^{-3}$, and $1 \cdot 2 \times 10^{-2}$ sec, respectively. The relatively very long diffusion time in D_2O (0·15 sec) is, of course, due to the extremely low thermal-neutron capture cross sections of deuterium (0·00057 barn) and oxygen (<0·0002 barn).

Although H_2O, D_2O, Be, and C are all useful moderating materials, in practice one finds that H_2O is the most widely used one. This is due to a combination of properties that make it attractive for this purpose:

(1) its low cost, which enables one to use large quantities of it, in an open pool, using it not only as moderator, but also as coolant and biological shield,

(2) its optical transparency, which allows one to observe a reactor core, with all the adjunct equipment placed near the core, at all times.

The earliest research reactors built were air-cooled assemblies of rectangular blocks of graphite and natural uranium. This was due to the fact that one can attain criticality with natural uranium if graphite (or D_2O) is used as moderator—but not if H_2O is used (because of its relatively high thermal-neutron capture cross section). If H_2O is used as moderator, one must have at least slightly-enriched uranium as fuel [natural uranium is only 0·720 atomic % ^{235}U, the remainder being almost all ^{238}U (99·274 atomic %), plus a little ^{234}U (0·0056 atomic %)]. The use of a pile of uranium and graphite blocks led to the term, "pile", that was first used. With the advent of other types of reactors, the term was replaced by the more general one, "nuclear reactor". The use of a core containing enriched uranium, immersed in a pool of H_2O as moderator/coolant/shield, led to the terms, "swimming pool reactor" and "pool reactor". Modern pool-type research reactors cost in the range of $200,000 to $1,000,000, depending upon the reactor type, whether above-ground or below-ground, the designed power level, and the experimental facilities built into them.

Whereas power reactors are designed for operation at very high temperatures (in order to attain high thermodynamic efficiencies for the conversion of thermal energy, generated by the fission reaction, to electrical energy), research reactors are usually designed to operate at close to room temperature (to avoid undesirable effects upon samples under investigation, when they are placed in or near the reactor core, i.e. effects such as fusion, vaporization, and decomposition). Thus, H_2O or D_2O, in liquid form, are attractive as coolants for research reactors. To be a good coolant, a liquid with suitable heat-transfer properties, and with a low absorption cross section for thermal neutrons, is desired. Otherwise, it need not necessarily be a good moderator for fast neutrons, but if it is, this is even more desirable. Both light water and heavy water are good in these respects.

In some types of research reactors, such as some models of the TRIGA reactor, the core, although immersed in a deep pool of water, is surrounded by an annular ring of graphite. The graphite improves the neutron economy, since it does not absorb thermal neutrons as much as would an equal volume of H_2O. Thus, many thermal neutrons that would otherwise be lost are scattered back into the core. High-purity beryllium (or BeO) or D_2O can also be used for this purpose, but they are considerably more expensive than even reactor-grade (high-purity) graphite.

In a below-ground pool reactor, where the earth is used as the main shielding in the lateral (horizontal) direction, the vertical depth of water required to give adequate shielding to personnel operating close to the surface of the pool depends, of course, upon the power level of operation of the reactor—since the neutron and γ-ray output of a reactor core is directly proportional to the power level of operation (since this is directly related to the rate of fissioning). Rather approximately, the water depth used, in practice, to reduce the dose rate at the pool surface to about 10 mr/hr is about 14 ft for 10 kW operation, 16 ft for 100 kW, and 18–20 ft for 1,000 kW (1 MW). Much of the measured radiation at the surface of a pool reactor is not due to thermal neutrons, fast neutrons, or γ-radiation coming from the reactor core, but rather consists of β- and γ-radiation coming from radionuclides such as 7·14-sec ^{16}N, 2·05-min ^{15}O, 29·1-sec ^{19}O, 9·96-min ^{13}N, 2·25-sec ^{15}C, and 110-min ^{41}Ar, generated in the water in and near the core, and rising to the surface by convection and in radiolysis gas bubbles. Their contribution can be greatly reduced by use of a jet pump diffuser system, mounted well down in the pool, which swirls the water around, in a horizontal plane, preventing these induced activities from rising to the surface so rapidly. Such a pump is used in TRIGA reactors. Since the major activity is ^{16}N, formed by the ^{16}O$(n, p)^{16}$N reaction, this enforced longer residence time markedly cuts down the surface radiation level, as ^{16}N has a half life of only 7·14 sec. Some of the surface γ-radiation consists of neutron-capture γ-rays, mostly from the ^{1}H$(n, \gamma)^{2}$H reaction. The ^{13}N arises from the ^{14}N$(n, 2n)^{13}$N reaction on dissolved N_2 in the water, and the ^{41}Ar arises from the ^{40}Ar$(n, \gamma)^{41}$Ar reaction on dissolved argon in the water (since air is 78·08% v N_2 and 0·934% v Ar). Air monitors are usually employed in the reactor area, and are set to sound an alarm if the radiation level goes above a preset safety level—possibly indicating a ruptured or leaking fuel element, or some other type of reactor malfunction.

At relatively low power levels of operation, natural convection cooling is sufficient, in pool reactors. For operation at higher power levels, forced convection plus an external circulation loop, with a heat exchanger, is needed. Usually, a small side stream of the water passes continually through an ion-exchange column. This removes any metallic and non-metallic ions from the water—resulting from impurities in the water, external contamination getting into the water, corrosion products, or fuel-element leakage. A γ-ray detector monitors the column, and sets off an alarm if there is any rapid increase in activity—possibly denoting a leaking fuel element.

D. Critical Mass, Reactor Control and Fuel Elements

When an appropriate assembly is made of a small isotopic neutron source, reactor fuel (fissionable material such as ^{235}U, in some suitable physical-chemical form), several control rods (containing appreciable amounts of high cross-section material, such as boron or cadmium), and a moderator material, by adding fuel to the assembly in increments, a point will be reached where the neutron production rate will slightly exceed the rate of loss of neutrons by escape and capture (assuming that the system is capable of becoming critical). In order to bring the now slightly supercritical system up to some desired power level of operation (and hence neutron-flux level), the power level, as measured by suitable neutron-monitor instruments located near the core, is allowed to rise almost exponentially until the desired power level is reached. The rate of power rise resulting from the branching-chain fission reaction is observed and recorded on a logarithmic "period meter" (the period of the reactor power increase rate is the time required to increase the power by a factor of e). At this point, one or more of the control rods is driven part way into the core until a steady state is attained; a state in which the rate of production of neutrons by fission at that power level just equals the rate by loss of neutrons (by escape, plus capture).

The mass of ^{235}U needed to just make a given system operate at zero power, with the neutron source removed, is referred to as the "critical mass" of the system. In pool reactors using 20% enriched uranium fuel and water moderator, this may typically amount to only 2–4 kg of ^{235}U. In a homogeneous solution ("water boiler") type of reactor, it may only amount to 0·6–0·8 kg ^{235}U (using a 90% enrichment). The water boiler type of reactor is limited to rather low powers (< 100 kW), and is subject to corrosion problems and gas bubble and radiolysis problems. In any given system, the exact value of the critical mass will depend upon many factors, such as the geometrical arrangement of the fuel, moderator, reflector, and coolant; the degree of enrichment of the fuel; and the absorption of neutrons by all the various materials used in and near the core.

Once a given reactor has been originally brought up to critically, subsequent operation and shutdowns are quite simple. When not being operated, the control rods are inserted deeply into the core, and the system is subcritical. When operation is desired, (1) the control rods are slowly withdrawn until the reactor is slightly supercritical, and (2) the power level is allowed to continue to rise at a moderate and safe rate until it has reached the desired level, when the control rods are partially re-inserted—just enough to maintain the desired steady-state condition. With modern research reactors, the entire startup procedure to desired operating level only requires a few minutes. Shutdown is achieved by "scramming" the reactor, i.e. rapidly re-inserting the control rods, making the system subcritical. At power levels above about 10 kW, the water in the vicinity of the core of a pool reactor is observed to emit a blue glow—the Cerenkov radiation, produced largely by Compton electrons generated in the water by Compton-scattering interactions of the γ-rays. The intensity of the Cerenkov blue glow increases with increasing power level.

Most nuclear reactors, including pool-type research reactors, have a small negative temperature coefficient (delayed, in reactors using plate-type fuel elements; prompt, in the TRIGA type of reactor, which uses cylindrical, hydrided, fuel elements). Thus, as a reactor warms up with operation, the control rods must be withdrawn slightly further, to maintain criticality. Most reactors are heavily instrumented for safety, with several independent automatic sensing and scramming devices— to shut down the reactor rapidly, should it begin to get out of control. As a further precaution, most reactors are placed in "confinement" buildings—to prevent the escape of dangerous, highly-radioactive fission products (gaseous and airborne particulate) into adjacent areas, in the event of a fuel-element rupture. Because of earlier concern over such possibilities, many of the older reactors were also located in rather isolated, little-populated regions. Particularly with the advent of intrinsically-safe reactors of the TRIGA type, remote siting and confinement buildings are no longer necessary.

The fuel elements used in research reactors of the pool type are usually either in the form of thin metallic plates (MTR type) or cylindrical hydrided metallic rods (TRIGA type), clad in aluminium, stainless steel, or zircalloy. Usually, the number of fuel elements employed is in the range of 25–100, depending upon the reactor design. Most pool-type research reactors employ 20%-enriched ^{235}U, alloyed or dispersed in a suitable metallic matrix such as zirconium, and water as the moderator and coolant. As mentioned earlier, ordinary uranium (0·720 atomic % ^{235}U) will not reach criticality in ordinary water (but will in D_2O or graphite).

The maximum steady power level (and hence steady neutronflux level) at which a given thermal-neutron nuclear reactor can be operated depends mainly upon the provisions for heat removal. In reactor designing, sufficient excess reactivity must be included to compensate for xenon poisoning. Xenon poisoning arises from the formation of ^{135}Xe, a fission product. This radionuclide has a half life of 9·13 hr, and is the third nuclide in the ^{135}Te fission-product decay chain:

$$^{135}\text{Te} \xrightarrow[\beta^-]{\sim 2m} {}^{135}\text{I} \xrightarrow[\beta^-]{6\cdot 68h} {}^{135}\text{Xe} \xrightarrow[\beta^-]{9\cdot 13h} {}^{135}\text{Cs} \xrightarrow[\beta^-]{3\cdot 0 \times 10^6 y} {}^{135}\text{Ba(stable)}.$$

Xenon-135 is unusual in that it has an extremely high thermal-neutron (n, γ) cross section: $3\cdot 5 \times 10^6$ barns. If a reactor is run for many hours, the ^{135}Xe poisoning will rise and then eventually reach a steady state, where its rate of formation from ^{135}I equals its rate of disappearance by (n, γ) reaction and β^- decay. This poisoning can be compensated for by gradual removal of control rods. However, when the reactor is shut down, the ^{135}Xe level in the fuel elements will continue to rise for about 10 hr, since it is still being formed by decay of accumulated ^{135}I (half life, 6·68 hr), but is now disappearing only by β^- decay (with a half life of 9·13 hr). With reactors operated at thermal-neutron fluxes of about 10^{13} n cm^{-2} sec^{-1} or less, the xenon poisoning does not cause any difficulty in re-startup of the reactor after a shutdown. However, with reactors operated at fluxes in the 10^{14}–10^{15} n cm^{-2}-sec^{-1} range, re-startup may not be possible for many hours (or even a day or two) after a shutdown, until the ^{135}Xe has decayed away sufficiently, unless adequate excess reactivity has been provided.

The next most severe fission-product, high cross-section, reactor poison is ^{149}Sm, the stable nuclide end product of the ^{149}Nd fission-product decay chain:

$$^{149}\text{Nd} \xrightarrow[\beta^-]{2\cdot 0h} {}^{149}\text{Pm} \xrightarrow[\beta^-]{54h} {}^{149}\text{Sm (stable)}.$$

However, its effect is relatively minor, as compared with that of ^{135}Xe. The thermal-neutron (n, γ) cross section of ^{149}Sm is also quite large: $4\cdot08 \times 10^4$ barns. Samarium may be used deliberately as a "burnable poison", where, typically, samarium is incorporated in the fuel elements. As the reactor is used over a long period of time, the buildup of high cross-section fission products in the fuel elements and consumption of fuel—resulting in a decrease in the reactivity of the core—is offset by consumption of ^{149}Sm in the fuel. The (n, γ) product, ^{150}Sm, has a relatively low (n, γ) cross section.

With research reactors operating intermittently, and at thermal-neutron fluxes of about 10^{13} n cm^{-2} sec^{-1} or less, fuel-element replacement is a rather minor problem. Under these conditions, the core lifetime is long (>10 years), and is primarily controlled by considerations of corrosion of the fuel-element cladding material. If a 10^{13} flux reactor is run almost continuously, the fuel-element replacement rate will be somewhat more extensive, of course, and at a 10^{14} flux, still more extensive. Operation at a power level of 1 MW corresponds to a fission rate of $3\cdot1 \times 10^{16}$ ^{235}U nuclei per second, or $1\cdot1 \times 10^{23}$ per 1,000 hr. This corresponds to 44 g of ^{235}U consumed per 1,000 hr (2·2% of 2 kg of ^{235}U). Non-fission neutron capture in the ^{235}U increases this to a 52 g loss of ^{235}U (2·6% of 2 kg). This is slightly offset by the formation of fissionable ^{239}Pu from ^{238}U. When fission-product buildup (poisoning) and fuel consumption in the fuel elements has approached the point where withdrawal of the control rods can barely bring the reactor to power, replacement of one or more of the older fuel elements is needed. This can be accomplished quite rapidly, in a pool reactor. The reactor is shut down; an element is removed, under water, with long-handled tools; a lead storage cask is lowered into the pool from an overhead crane; the highly-radioactive spent fuel element is placed in the cask; and the cask is capped and brought to the surface for later disposal (sometimes such spent fuel elements are placed in a deep pool of water and used as a high-intensity γ-ray source for radiation chemistry studies). The new fuel element is then inserted in the reactor core lattice, in the position just vacated, and the reactor is ready for use again.

E. Reactor Neutron Fluxes and Flux-Monitoring Techniques

As discussed earlier, the typical research-type thermal-neutron reactor requires high fluxes of thermal neutrons in the core, to maintain operation at a high power level, and generates high fluxes of fast neutrons by the fission reaction. Thus, within and near the core of such a reactor there exist high fluxes of neutrons ranging in energy from thermal energies (0·025 eV average at 17°C) all the way up to many MeV. The neutron flux distribution at any particular point in such a system will depend upon many factors.

For any given system, the average thermal-neutron flux within the core will be directly proportional to the power level of operation. The proportionality constant includes several system constants, one of which is the critical mass of the system. In general, the thermal-neutron flux will be higher than this average value, near the center of the core, and lower than the average value, near the periphery of the core. The same trend is true for the fast-neutron flux: highest near the center of the core, least near the periphery. Furthermore, the ratio of fast-neutron flux to thermal-neutron flux increases in going from the periphery of the core in toward the center. These points are illustrated, for one particular reactor, a 250 kW TRIGA Mark I reactor (with annular graphite reflector; water-moderated and cooled), in Table I. This reactor consists of a cylindrical array of about 70 fuel elements, arranged in concentric rings. The center hole is called the "A ring", followed by the B, C, D, E, and F rings of fuel elements, the F ring, being the outermost ring of the core. Each element consists of a 15 in high length of 1·4 in diameter zirconium (91·5%)—uranium (8·5%) alloy that has been hydrided to a 1/1 Zr/H atomic ratio. The uranium is

20% enriched (20% ^{235}U, 80% ^{238}U). At each end of the fuel rod, there is a 4 in length of graphite, and the entire fuel element is clad in aluminium, making it water-tight. The overall core is 18 in in diameter by 28 in high. Three electric motor-driven boron control rods provide the means for startup, steady operation at any desired power level up to its licensed maximum level of 250 kW, and shutdown. A fourth control rod, pneumatically operated, is used for pulse operation. The core is located in a pool of de-ionized water 6·5 ft in diameter by 20 ft deep (16 ft of water above the top of the core). An outside loop constantly runs the pool water through a heat exchanger and an ion-exchange column.

TABLE I

TRIGA Mark I Neutron Fluxes for Steady-State Operation at 250 kW

Irradiation Position	Neutron Flux (n cm^{-2} sec^{-1})				
	Thermal	>10 keV	>1·35 MeV	>3·68 MeV	>6·1 MeV
Rotary specimen rack ("lazy susan")	1·8 × 10^{12}	1·5 × 10^{12}	1·8 × 10^{11}	2·5 × 10^{10}	4·0 × 10^{9}
Fuel-element ring D	4·9 × 10^{12}	9·0 × 10^{12}	2·3 × 10^{12}	4·1 × 10^{11}	6·2 × 10^{10}
Fuel-element ring F ("rabbit" terminus)	4·3 × 10^{12}	3·5 × 10^{12}	7·5 × 10^{11}	1·25 × 10^{11}	1·9 × 10^{10}
Reactor pool outside reflector	6·8 × 10^{11}	6·8 × 10^{10}	5·1 × 10^{9}	6·3 × 10^{8}	1·1 × 10^{8}

For most high-flux, high-sensitivity neutron activation analysis work, a high thermal-neutron flux is desired, as free as possible of epithermal and fast neutrons. Unfortunately, it is not possible to produce a really high, purely thermal, neutron flux. The thermal-neutron flux, as mentioned earlier, is highest at or near the center of the core—where, however, the fast-neutron flux is also highest. As one moves out of the core into the surrounding moderator, the thermal-to-fast ratio improves, but the thermal- and fast-neutron fluxes both drop off. As can be seen from the flux values in Table I, as one proceeds (in the 250 kW TRIGA Mark I reactor) from the D ring, to the F ring, to the rotary specimen rack (inside the graphite reflector), to the pool just outside the graphite reflector, the ratio of the thermal-neutron flux to the flux of neutrons above 1·35 MeV improves from 1·2 (to 1), to 5·7, to 10, to 133. However, in the same sequence of locations, the thermal-neutron flux decreases by a factor of 7·2, in going from the D ring to the pool region just outside the graphite reflector.

Light water is a good moderator in that it has a very short slowing down length (5·74 cm for 2·0 MeV to 0·025 eV), but suffers from the appreciable thermal-neutron capture cross section of ^{1}H (0·33 barn), which results in a very short diffusion length (2·88 cm) and migration length (6·43 cm). Heavy water, Be, and graphite all have greater slowing down lengths (10·9 cm, 9·90 cm, and 18·7 cm, respectively), but compensate by not capturing the thermalized neutrons so efficiently—resulting in longer diffusion lengths (100 cm, 23·6 cm, and 50·2 cm, respectively) and longer migration lengths (101 cm, 25·8 cm, and 53·6 cm, respectively) than H_2O. Thus, these moderators can provide a higher thermal-to-fast ratio than H_2O, but, require a greater thickness of moderator (because of their appreciably longer slowing down lengths). At such increased distances from the center of the core, all the neutron fluxes decrease somewhat, because of the quasi inverse-square effect. Where the highest thermal-to-fast ratios are desired, a "thermal column" consisting of a thick block of graphite or a tank filled with D_2O, placed close to the periphery of the core, is used. If samples are placed at the outer end of the thermal column, or inserted into the column at some point along the length of the column, they will be exposed to a considerably higher thermal-to-fast ratio than if they were placed just at the periphery of the core—or further out in H_2O.

Implicit in the above discussion of neutron fluxes of various energies, and the listing of illustrative fluxes of various neutron energies in Table I, is the assumption that means exist for the measurement of such fluxes. This assumption is quite valid, as various neutron-flux measurement techniques do exist, and are widely used. However, in general, it should be remarked that the accurate measurement of the flux of neutrons of a specified narrow energy range—in a neutron flux consisting of a broad spectrum of energies—is not a simple matter. A rigorous and detailed treatment of the subject of neutron flux measurement techniques would be very lengthy, and is beyond the scope of this chapter. Instead, a brief synopsis of the subject is given below. Fortunately, the neutron activation analyst has little need for a really detailed and comprehensive knowledge of the subject. This is because he normally employs a comparator, rather than absolute, technique. Conducted properly, the comparator technique eliminates the need for a detailed knowledge of the neutron flux-energy distribution in the region of the reactor where he places his samples and comparator samples.

The ideal thermal-neutron flux-monitor element, or separated stable nuclide, is one that not only forms a readily-detected radionuclide product by (n, γ) reaction, but one whose (n, γ) cross section falls off only as $1/V$, with no (n, γ) resonances. No element is quite ideal, but the different elements vary considerably from one another in their approach to this ideal. Elements that have numerous high cross-section (n, γ) resonances are clearly not suitable for measuring the thermal-neutron flux. Many of the medium-Z to high-Z elements are thus not suitable. Resonances shown by low-Z elements are largely due to neutron scattering, rather than to radiative capture, and hence some of the lower-Z elements are more suitable as thermal-neutron flux monitors. In practice, quite a few different elements are used, by various workers, to measure thermal-neutron fluxes—to various degrees of accuracy, depending upon the element chosen, and the overall neutron-energy flux distribution. Manganese, for example, is frequently used—via the ^{55}Mn$(n, \gamma)^{56}$Mn reaction ($\sigma = 13 \cdot 3$ barns), and detection of the 0·847 MeV γ-rays of 2·58-hr ^{56}Mn. Manganese-55 does exhibit some strong resonances in the epithermal neutron-energy region, but these are largely due to neutron scattering, rather than to (n, γ) reaction to form ^{56}Mn. In addition to manganese, Na, Al, and Cu are often used as thermal-neutron flux monitors. These elements have thermal-neutron isotopic (n, γ) cross sections of 0·525 barn, 0·21 barn, and 4·51 barns, respectively, to form 15·0-hr ^{24}Na, 2·31-min ^{28}Al, and 12·8-hr ^{64}Cu. Sodium and aluminium are monoisotopic, whereas copper is 69·09 atomic % ^{63}Cu. The principal γ-ray energies of these three radionuclides are 1·37 MeV (and 2·75 MeV), 1·78 MeV, and 0·511 MeV (from β^+ annihilation), respectively. With a known weight of monitor element, of known (n, γ) cross section, exposed for a known period of time, and then counted at a known counting efficiency, it is possible to calculate the effective thermal-neutron flux in that irradiation position. Since the sample is exposed to a whole spectrum of neutron energies, and the (n, γ) cross section in the low-energy region (say, below 0·2 eV) declines only as 1/V, the "thermal-neutron flux" value so obtained is really an effective, or integrated average value. Thus, slightly different values may be obtained with different monitor elements.

In the epithermal neutron-energy region (which can be arbitrarily defined as the region from 0·2 eV to 1,000 keV), many elements—particularly the higher-Z elements—exhibit a number of strong (n, γ) resonances, i.e. narrow energy regions in which the (n, γ) cross section rises far above the 1/V level, often to very high values. For this reason, this region of the neutron-energy spectrum is often called the "resonance region", and activation of nuclides by neutrons in this energy region is called "resonance activation". In this region, the resonance neutron flux per unit energy is approximately proportional to $1/E_n$. The resonance flux can be measured by activation of a suitable monitor element with, and without, cadmium shielding. Without the cadmium, the element is activated by both thermal and resonance (epithermal) neutrons; with cadmium, the thermal-neutron activation (up to about 0·4 eV) is essentially eliminated, while the resonance activation is virtually unaffected. Three elements commonly used to measure the resonance neutron flux are cobalt, indium, and gold. These elements have thermal-neutron (n, γ) cross sections of 37 barns, 145 barns, and 99 barns, respectively, and "resonance integral" (n, γ) cross sections of 75 barns, 2635 barns, and 1558 barns, respectively, to form 5·26-yr ^{60}Co, 54·0-min ^{116m}In, and 2·70-dy ^{198}Au. The principal γ-rays emitted by these three radionuclides are 1·17 MeV (and 1·33 MeV), 1·29 MeV, and 0·412 MeV, respectively. Cobalt and gold are monoisotopic, whereas indium is 4·28 atomic % ^{113}In and 95·72 atomic % ^{115}In.

To measure the fast-neutron flux, one employs "threshold" detector elements, i.e. elements that undergo an endoergic fast-neutron reaction, with a known threshold energy. One can select detector elements with reaction thresholds of anywhere from tenths of an MeV to even 25 MeV, as desired. For example, some nuclides undergo an (n, n') reaction, to form a metastable isomeric state whose subsequent isomeric-transition decay can be detected. Some of these (n, n') detector nuclides have thresholds in the range of tenths of an MeV. Nuclides that undergo endoergic (n, p) or (n, α) activation, to form a suitably-detectable radionuclide product, in many cases have thresholds in the range of 1–5 MeV; a few even higher—for example, the ^{16}O$(n, p)^{16}$N reaction has a threshold of 10·2 MeV. The ^{32}S$(n, p)^{32}$P reaction is frequently used. This reaction has a Q value, calculated from the atomic masses, of $-0·926$ MeV, i.e. a threshold energy of $0·926 \times 33/32$, or 0·955 MeV. The (n, p) excitation curve for ^{32}S thus shows a zero cross section for neutron energies below 0·955 MeV, since the reaction is not thermodynamically possible at neutron energies below this value. Above 0·955 MeV, the cross section for the reaction rises only slowly until neutron energies equal to, or greater than the classical Coulomb barrier to escape of the proton from the ^{33}S* compound nucleus are reached. The barrier in this example, calculating the barrier height from the equations, $B = (Ze)(ze)/R$, $R = r_p + r_{32P}$ and $r = 1·5 \times 10^{-13}$ A$^{1/3}$ cm, is 3·44 MeV. Competition from decay of the compound nucleus via other reaction paths, such as the ^{32}S$(n, \alpha)^{29}$Si and ^{32}S$(n, 2n)^{31}$S reactions (which have Q values of $+1·52$ MeV and $-15·07$ MeV, respectively) influence the shape of the ^{32}S$(n, p)^{32}$P excitation function above its threshold.

Threshold detectors thus provide some measure of the neutron flux above the various thresholds. However, since the neutron flux in the range of about 0·1 MeV to around 2 MeV, in any core or near-core position in a reactor, has a shape that is not accurately known, even the use of several different threshold detectors having thresholds within this range, and having known excitation curves, does not provide a really accurate representation of the fast-neutron flux in this energy region. Above about 2 or 3 MeV, the neutron flux in any given region of a reactor more closely approximates a fission-spectrum neutron flux, and hence the use of various high-threshold detectors with known excitation functions does give a fairly accurate representation of the fluxes of higher-energy neutrons. The fast-neutron fluxes shown in Table I, for neutrons above 1·35 MeV, above 3·38 MeV, and above 6·1 MeV (in various locations in the 250 kW TRIGA Mark I reactor) were determined by the use of various threshold detectors.

F. Irradiation of Samples in Research Reactors in Neutron Activation Analysis Studies

In neutron activation analysis measurements, using a research reactor, one usually wishes to expose one or more samples and standards to a highly thermalized neutron flux. In some instances, however, a predominantly fast-neutron flux is desired. In either case, the desired irradiation times can range, in practice, from fractions of a second on up to many hours. Ways and means of conducting such irradiations readily and reproducibly constitute the subject matter of this section.

1. THERMAL-NEUTRON ACTIVATIONS

If one desires to minimize the occurrence of possible fast-neutron reactions in the samples to be activated, one normally chooses an irradiation position outside the core of the reactor—in the surrounding moderator/reflector material (H_2O, D_2O, Be, or graphite). Typically, as illustrated in Table I, the fast-neutron flux in such out-of-core positions, although reduced relative to the thermal-neutron flux, may still be appreciable. For example, samples rich in sulfur and/or chlorine may still produce significant amounts of ^{32}P, via the $^{32}S(n, p)^{32}P$ and $^{35}Cl(n, \alpha)^{32}P$ reactions, respectively. If one were analysing such samples for relatively low levels of phosphorus, via the $^{31}P(n, \gamma)^{32}P$ reaction, one would obtain erroneously high phosphorus values unless appropriate corrections were made for the ^{32}P formed by the S and/or Cl present. There are numerous other practical examples where erroneously high values can be obtained for one element of interest, if the samples contain higher levels of the elements one or two units higher in Z, if these can form the same radionuclide, via (n, p) and (n, α) reactions, respectively, that the element of interest forms via the (n, γ) reaction.

A few additional examples of practical interest may be cited here: formation of ^{20}F not only from fluorine, but also via the ^{23}Na$(n, \alpha)^{20}$F reaction; formation of ^{27}Mg not only from magnesium, but also via the ^{27}Al$(n, p)^{27}$Mg and ^{30}Si$(n, \alpha)^{27}$Mg reactions; formation of ^{38}Cl not only from chlorine, but also via the ^{41}K$(n, \alpha)^{38}$Cl reaction; formation of ^{51}Ti not only from titanium, but also via the ^{51}V $(n, p)^{51}$Ti and ^{54}Cr$(n, \alpha)^{51}$Ti reactions; formation of ^{56}Mn not only from manganese, but also via the ^{56}Fe $(n, p)^{56}$Mn and ^{59}Co $(n, \alpha)^{56}$Mn reactions, etc.

In practice, one corrects for such possible fast-neutron reaction product interferences by activating samples with, and without, a surrounding layer of cadmium. Due to the very high thermal-neutron absorption cross section of cadmium (2537 barns), even a very thin layer of cadmium will absorb almost all the thermal neutrons incident upon it. For example, a layer of cadmium only 1 mm thick will decrease the transmitted thermal-neutron flux to a level only about one millionth of the incident flux. The use of such a thin layer of cadmium does not appreciably alter the incident fast-neutron flux, as the neutron-absorption cross section of cadmium for fast neutrons is quite low. Hence, for example, a sample containing a relatively large amount of sodium, relative to the amount of fluorine in the sample, would produce 11·6-sec ^{20}F from both the ^{19}F$(n, \gamma)^{20}$F reaction, and the ^{23}Na$(n, \alpha)^{20}$F reaction—in a mixed thermal-/fast-neutron flux. With a cadmium shield around the sample, the ^{20}F production would be reduced, since its formation from fluorine would be greatly reduced—its formation from sodium unchanged. Thus, one might expect the relationships:

Unshielded: (cps ^{20}F)$_0$ = a(wt F)+b(wt Na).

Cd-Shielded: (cps ^{20}F)$'_0$ = b(wt Na).

Then, (cps ^{20}F)$_0$−(cps ^{20}F)$'_0$ = a(wt F).

The constants, a and b, are the ^{20}F photopeak counting-rate specific activities (under prescribed counting conditions) for ^{19}F→^{20}F and ^{23}Na→^{20}F, respectively, each expressed in (cps ^{20}F)$_0$ per gram of element (fluorine or sodium), individually determined experimentally for that irradiation position by the activation of fluorine and sodium standards.

This is almost the case, but not quite. Although a thin layer of cadmium will remove thermal neutrons almost quantitatively, it does not absorb epithermal neutrons so effectively. Many of these epithermal neutrons will therefore reach the sample, in spite of the cadmium, and they can also produce (n, γ) activation of the fluorine. The cross section for the (n, γ) reaction with most nuclides is lower for epithermal neutrons than for thermal neutrons, since for most nuclides the (n, γ) cross section falls off approximately as 1/V, where V is the neutron velocity. However, for many nuclides, there are a number of (n, γ) resonances. Hence, super-imposed upon the 1/V relationship, there are a number of peaks in the (n, γ) cross section—versus—neutron energy curve, so-called "resonance-absorption peaks".

As a result of activation by epithermal-resonance neutrons, the (n, γ) activation of a stable nuclide is not reduced to almost zero by a 1 mm thickness of cadmium. Instead, depending upon the incident neutron energy-flux distribution and the particular nuclide, it will typically only be reduced by a factor of between perhaps 5 and 20. This ratio of (n, γ) activation, without and with an appreciable layer of cadmium, is called the "cadmium ratio" for that particular irradiation position—and for that particular stable nuclide.

Thus, for the fluorine-sodium example cited, the more exact relationships are the following:

Unshielded: (cps ^{20}F)$_0$ = a(wt F)+b(wt Na)

Cd-Shielded: (cps ^{20}F)$'_0$ = a'(wt F)+b(wt Na),

in which a' is the specific yield of ^{20}F from ^{19}F, when shielded by cadmium, and a/a' is the cadmium ratio for that irradiation position, for the (n, γ) activation of ^{19}F. Thus, the two equations can be solved for the unknown, (wt F). Fluorine-20 can also be formed by the ^{20}Ne$(n, p)^{20}$F reaction, but this reaction was excluded from consideration in this example, since neon, being an inert gas, would not normally be present in solid or liquid samples to any significant extent. Actually, ^{19}F is appreciably activated by epithermal neutrons, by (n, γ) reaction, but does not exhibit (n, γ) resonances.

There are other cases, however, such as some of those mentioned earlier in this section, where one must consider all three reactions: (n, γ), (n, p), and (n, α)—if one wishes to determine one of the elements formed by fast-neutron reaction. In the case of the determination of low levels of phosphorus, via the ^{31}P$(n, \gamma)^{32}$P reaction, in matrices that may contain higher levels of sulfur, and/or chlorine, one must correct for the production of ^{32}P also by the ^{32}S$(n, p)^{32}$P and ^{35}Cl$(n, \alpha)^{32}$P reactions. Thus,

Unshielded: (cps ^{32}P)$_0$ = a (wt P)+b(wt S)+c(wt Cl)

Cd-Shielded: (cps ^{32}P)$'_0$ = a'(wt P)+b(wt S)+c(wt Cl).

From the cadmium ratio, a/a', determined for that position for the ^{31}P$(n, \gamma)^{32}$P reaction, using a phosphorus standard, these two equations can be solved directly for (wt P). However, if one wishes to solve for (wt S) and/or (wt Cl), a third independent equation, with no additional unknowns, is needed. In this particular case, one usually independently determines (wt Cl) via the $^{37}Cl(n, \gamma)^{38}Cl$ reaction, thus providing the third equation. (Technically, the ^{38}Cl production should be corrected for possible concurrent production via the $^{38}Ar(n, p)^{38}Cl$ and $^{41}K(n, \alpha)^{38}Cl$ reactions. However, even if some air were present in the sample vial, air is only $0.934\%v$ argon, and argon is only 0.063 atom% ^{38}Ar, so this source of ^{38}Cl is negligible. The yield of ^{38}Cl from potassium is only significant in samples where the K/Cl ratio is quite large.)

In some reactor irradiation positions, the assumption that the fast-neutron specific yields (b and c in the equations) are the same, whether or not cadmium is used, can be somewhat inaccurate. The use of an appreciable amount of cadmium near one or more fuel elements may cause a neutron flux depression in that region that affects the fast-neutron flux as well as the thermal-neutron flux, incident on the outside surfaces of the cadmium. If such is the case, the b and c in the cadmium-shielded equation must be replaced by b' and c', determined experimentally by irradiation of standards with cadmium. Experimentally, it has been shown that this problem is not of any consequence in irradiations carried out with cadmium shielding in the rotary specimen rack of the author's TRIGA Mark I reactor, but can be of importance in pneumatic-tube irradiations—where the sample is much closer to the fuel elements.

H. R. Lukens has compiled an excellent table of calculated yields of thermal-neutron and fast-neutron reaction products in reactor irradiations (Lukens, 1964). This compilation gives the approximate photopeak counting rate for each γ-ray or positron emitter formed, via (n, γ), (n, n'), (n, p), (n, α), and $(n, 2n)$ reactions, from all the elements—some 896 reactions in all. The (n, γ) cross sections used were those from the compilation by Hughes and Schwartz (1958); the (n, p), (n, α), and $(n, 2n)$ cross sections, for a fission neutron spectrum, were those from the compilation by Roy and Hawton (1960). These are calculated for a 1-hr irradiation in a neutron flux consisting of $3.5 \times 10^{12} n$ cm^{-2} sec^{-1} (thermal) and 3.5×10^{12} n cm^{-2} sec^{-1} (fission-spectrum). The photopeak counting rates are for the major γ-ray produced, for counting with a 3×3 in NaI(Tl) detector, with a 2 cm distance between sample and crystal (using the detector photopeak efficiency tabulation of Heath, 1964), and for zero decay time. Half lives and decay schemes were taken from the compilation by Strominger et al. (1958). For example, Luken's compilation indicates that, for these conditions, the ^{51}Ti photopeak counting rate (0.320 MeV) would be about 2.8×10^9 cpm/g Ti, by (n, γ) reaction on ^{50}Ti, 1.8×10^8 cpm/g V, by (n, p) reaction on ^{51}V, and 4.0×10^6 cpm/g Cr, by (n, α) reaction on ^{54}Cr. Thus, under these conditions, the correction for fast-neutron produced ^{51}Ti would amount to 5% of the ^{50}Ti(n, γ) ^{51}Ti value if the V/Ti weight ratio in the sample were at least about 0·78/1, or if the Cr/Ti weight ratio were at least about 35/1. The values listed in this compilation are useful guides to the approximate extent of fast-neutron product interferences, but they should not, of course, be used in actual analytical work—experimentally determined values, obtained under the exact flux conditions involved in the sample irradiations, should instead be used. Also, it should be noted, one usually employs an irradiation position in which the thermal-neutron flux is appreciably larger than the fission-spectrum neutron flux (rather than a position in which they are of equal size)—if one is primarily interested in detecting an (n, γ) product. For example, in the rotary specimen rack of the TRIGA Mark I reactor, the thermal-neutron flux is about 10 times larger than the fission-spectrum neutron flux; in the pneumatic-tube position, this ratio is about 5·7/1.

It should also be remarked that some neutron-induced activities are formed concurrently, from the same element, by (n, γ) and $(n, 2n)$ reactions, or sometimes from (n, γ), $(n, 2n)$, and (n, n') reactions. Thus, for example, 120-dy ^{75}Se is formed concurrently by the ^{74}Se$(n, \gamma)^{75}$Se reaction and the ^{76}Se$(n, 2n)^{75}$Se reaction—and, 17·5-sec ^{77m}Se is formed concurrently by the ^{76}Se$(n,\gamma)^{77m}$Se, ^{77}Se$(n, n')^{77m}$Se and ^{78}Se$(n, 2n)^{77m}$Se reactions. Their relative contributions depend upon the neutron energy spectrum, the integrated cross sections of the individual reactions for this energy spectrum, and the abundances of the target stable nuclides. In a fairly well-thermalized neutron flux, the contribution from the $(n, 2n)$ reaction will usually be rather small, since such reactions typically are endoergic to the extent of about 8–12 Mev, and there are relatively few neutrons in such a neutron flux that are above 8–12 MeV in energy. The $(n, 2n)$ contribution, however, can be appreciable in a less-thermalized neutron flux, and especially if the (n, γ) cross section is rather low, and/or the isotopic abundance of the $(n, 2n)$ target nuclide is much greater than that of the (n, γ) target nuclide. For example, from Lukens' compilation it is seen that the ^{137m}Ba yields from the ^{136}Ba(n, γ) ^{137m}Ba and ^{138}Ba$(n, 2n)$ ^{137m}Ba reactions are in the ratio of only 1·6/1 (for equal thermal-neutron and fission-spectrum neutron fluxes). The isotopic abundance ratio of ^{136}Ba/^{138}Ba (7·81% to 71·66%) favors the relative yield from the $(n, 2n)$ reaction, as does the rather low (n, γ) cross section of ^{136}Ba (0·016 barn). Opposing this tendency is the relatively low flux of neutrons above about 10 MeV in energy and the generally low cross sections of $(n, 2n)$ reactions (in a fission-spectrum neutron flux, the cross section for the

^{138}Ba$(n, 2n)$ ^{137m}Ba reaction is only 1·1 millibarn) (Roy and Hawton, 1960).

In Table II, the interference-free limits of detection for 71 elements, in a thermal-neutron flux of 10^{13} n cm^{-2} sec^{-1}, and for a maximum irradiation time of one hour, are given. These are only for elements that form an (n, γ) product that emits γ-rays or positrons, and that has a half life longer than 5 sec. For each element that forms more than one (n, γ) gamma- or positron-emitting radionuclide, only the most sensitive one is listed. These values are largely taken from the tabulation of

TABLE II

Limits of Detection for 71 Elements in a Thermal-Neutron Flux of 10^{13} n cm^{-2} sec^{-1} (1 hr irradiation)

μg Limit of Detection	Elements
$1-3 \times 10^{-6}$	Dy
$4-9 \times 10^{-6}$	Mn
$1-3 \times 10^{-5}$	Kr, Rh, In, Eu, Ho, Lu
$4-9 \times 10^{-5}$	V, Ag, Cs, Sm, Hf, Ir, Au
$1-3 \times 10^{-4}$	Sc, Br, Y, Ba, W, Re, Os, U
$4-9 \times 10^{-4}$	Na, Al, Cu, Ga, As, Sr, Pd, I, La, Er
$1-3 \times 10^{-3}$	Co, Ge, Nb, Ru, Cd, Sb, Te, Xe, Nd, Yb, Pt, Hg
$4-9 \times 10^{-3}$	Ar, Mo, Pr, Gd
$1-3 \times 10^{-2}$	Mg, Cl, Ti, Zn, Se, Sn, Ce, Tm, Ta, Th
$4-9 \times 10^{-2}$	K, Ni, Rb
$1-3 \times 10^{-1}$	F, Ne, Ca, Cr, Zr, Tb
$4-9 \times 10^{-1}$	
$1-3$	
$4-9$	
$10-30$	Si, S, Fe

Lukens (1964), converted to a 10^{13} flux. The criteria for deriving a microgram limit of detection from the element specific activity (photopeak cpm/g of element at zero decay time, using a 3×3 in NaI(Tl) crystal with a sample-to-crystal distance of 2 cm) are those of Buchanan (1961) (minimum detectable photo-peak cpm of 1,000 for half lives < 1 min; 100 for half lives between 1 min and 1 hr; 10 for half lives greater than 1 hr). It is seen from the Table that these limits of detection range from as low as about 10^{-6} μg (for Dy) to as high as $10-30$ μg (for Si, S, and Fe), with a median value of about 0.001 μg. Thus, a great many elements can be determined down to extremely low levels: 45 down to levels of about $0 \cdot 001\,\mu g$ and lower; 59 down to $0 \cdot 01\,\mu g$ and lower; 65 down to $0 \cdot 1\,\mu g$ and lower. In a 1 g sample, these μg limits of detection correspond numerically to ppm limits of detection. In the absence of appreciable interferences from other induced activities, all of these limits of detection can be attained purely instrumentally. If interferences require it, these limits can still be attained by the use of post-irradiation radiochemical separations with carriers. Conducted properly, the method is strictly quantitative (to accuracies of ± 1–3% of the value) all the way from a 100% concentration down to levels approaching the limit of detection.

2. FAST-NEUTRON ACTIVATIONS

In the previous section, reactor fast-neutron activation products were discussed largely from the standpoint of their generation, in some instances, as unwanted interferences from elements of $Z+1$ and/or $Z+2$, via the (n, p) and (n, α) reactions, respectively, with the (n, γ) detection of an element of atomic number Z. However, in a number of instances the reactor fast neutrons can be put to good use. Some elements simply do not produce much activity, per unit weight, via (n, γ) reaction—due to either an unusually low (n, γ) cross section or a low isotopic abundance, or both—whereas a fast-neutron reaction gives a better yield of radioactive product. For example, the $^{14}N(n, 2n)^{13}N$ reaction produces an ^{13}N specific activity some 5,000 times greater than the ^{16}N specific activity produced by the $^{15}N(n, \gamma)^{16}N$ reaction— at equal thermal-neutron and fission-spectrum neutron fluxes. The principal reasons in this case are the exceptionally low (n, γ) cross section of ^{15}N (0·00002 barn), and the low isotopic abundance of ^{15}N (0·37% whereas nitrogen is 99·63% ^{14}N). Thus, fast-neutron activation of nitrogen is, under these conditions, much more sensitive than thermal-neutron activation. Similarly, detection of oxygen, via the $^{16}O(n, p)^{16}N$ reaction, is about 13 times more sensitive than detection via the $^{18}O(n, \gamma)^{19}O$ reaction, under these same conditions. This is largely due to the low (n, γ) cross section of ^{18}O(0·0002 barn), and the low isotopic abundance of ^{18}O(0·204%, whereas oxygen is 99·76% ^{16}O).

Other elements (besides nitrogen and oxygen) which form gamma emitters or positron emitters via thermal-neutron and fast-neutron reactions, for which a fast-neutron product provides more sensitive detection than does the highest-yield (n, γ) product, under these conditions, include silicon and iron. For fluorine, chromium, nickel, rubidium, and selenium, the limits are about the same with thermal neutrons and fast neutrons. There are also a few elements, such as phosphorus, thallium, and lead that yield only (or essentially only) a pure β^- emitter as the (n, γ) product, but which produce a useful yield of a γ emitter by a fast-neutron reaction. For some other elements, a fast-neutron product yield may be within a factor of 10 of the thermal-neutron product yield, and the fast-neutron product may have a half life and/or a γ-ray energy that make it more desirable as an indicator radionuclide than the thermal-neutron product—in the purely instrumental determination of the element in a matrix that generates one or more other (n, γ) products that interfere with the detection of the desired element.

TABLE III

Limits of Detection for 73 Elements in a Fission-Spectrum Neutron Flux of 10^{13} n cm^{-2} sec^{-1} (1 hr irradiation)

μg Limit	Elements
$1-3 \times 10^{-2}$	Al, Ni, Se, In
$4-9 \times 10^{-2}$	Si, Rb, Xe, Pr, Ir
$1-3 \times 10^{-1}$	F, Mg, Ti, V, Fe, Zn, Ga, As, Br, Kr, Sr, Y, Sb, Gd, Ho, Hf, Ta, W, Os, Au, Hg, Pb
$4-9 \times 10^{-1}$	P, K, Sc, Cr, Co, Te, Ba, Ce
1–3	N, Na, Ar, Ge, Zr, Nb, Mo, Ru, Rh, Pd, Ag Cd, Sn, Nd, Sm, Lu, Pt
4–9	Cs, Re
10–30	Ne, Cl, Cu, I, Eu, Tb, Er
40–90	Mn, Tm, Yb
100–300	O, Ca, La, Tl
400–900	S

Thus, there are a number of cases in which reactor fast-neutron activation can be of distinct use. In such instances, one usually needs to suppress the (n, γ) activation of other elements in the samples, so that their interference to the instrumental detection of the product of interest is minimized, or their complication of the radiochemical separation procedures is reduced. The (n, γ) suppression is readily accomplished by activation of the sample in a secondary container rich in cadmium or boron, as mentioned earlier.

Table III lists the best fission-spectrum limit of detection (via fast-neutron activation to form a gamma- or positron-emitter with a half life of greater than 5 sec) for 73 elements. These values are taken from the tabulation of Lukens (1964), converting to a fission-spectrum flux of 10^{13} n cm^{-2} sec^{-1}, and using the limit of detection criteria of Buchanan (Buchanan, 1961). Due mainly to the generally lower cross sections for fast neutrons, as compared with thermal-neutron (n, γ) cross sections, these limits of detection (with the exceptions of N, O, F, Si, P, Cr, Ni, Fe, Rb, Se, Tl, and Pb noted earlier) are generally considerably larger than those of Table II. Whereas amny isotopic thermal-neutron (n, γ) cross sections are of the order of 0·1 barn or larger (some being even as high as 10,000 barns), the fissoin-spectrum isotopic cross sections (to form the fast-neutron products giving the sensitivities listed in Table III) are much smaller—ranging in most cases from 0·5 to 5 millibarns for the (n, p) reactions, 0·1 to 1 millibarns for the (n, α) reactions, and 1–10 millibarns for the $(n, 2n)$ reactions. Largely because of the increase in Coulomb barrier with increasing product Z, an (n, p) product gives the best limit of detection (relative to other fast-neutron products) only for the lower Z elements, i.e. for 25 of the first 50 elements (when listed in order of increasing Z) of Table III (5 being better by (n, α) reaction; 20 being better by (n, n') or $(n, 2n)$ reaction). The remaining 23 elements of Table III (from Ce, $Z = 58$, on) are most sensitively determined by (n, n') or $(n, 2n)$ reaction, largely because these reactions have no Coulomb barrier—even though the $(n, 2n)$ reactions are all quite endoergic (7–11 MeV).

The fission-spectrum fast-neutron limits of detection shown in Table III range from about $0·01\,\mu g$ (Al, Ni, Se, In) up to about $500\,\mu g$ (S), with a median value of about $0·5\,\mu g$. This is about 500 times greater than the thermal-neutron median limit of detection for mostly the same elements, under comparable conditions of irradiation time (1 hr) and flux ($10^{13}\,n\,cm^{-2}\,sec^{-1}$ thermal and fission-spectrum), and counting conditions.

Some work has been done on the enhancement of fast-neutron fluxes by the use of a ^{235}U or LiD shield around samples, instead of Cd. Both ^{235}U and LiD are, like Cd, good absorbers of thermal neutrons, but, in addition, each generates more fast neutrons: ^{235}U by fission, and LiD by the $^6Li(n, \alpha)t$ reaction, followed by the $^2H(t, n)^4He$ reaction—to generate 14 MeV neutrons. The fast-neutron flux enhancements possible with the use of a ^{235}U or LiD annulus, however, do not appear to be very large.

3. PROVISIONS FOR IRRADIATION OF SAMPLES

For the reactor irradiation of samples up to nvt values of about $10^{17}\,n\,cm^{-2}$ (e.g. up to about a 1 hr irradiation at a thermal-neutron flux of $3 \times 10^{13}\,n\,cm^{-2}\,sec^{-1}$, or its equivalent) polyethylene is usually a satisfactory sample-container material. High-purity polyethylene vials of various shapes and sizes, ranging from internal volumes of about 0·1 cm³ up to 10 cm³, are widely used for such irradiations, so long as the sample-exposure temperatures do not greatly exceed room temperature. Even the best polyethylene available, however, usually contains measurable, though low, levels of several elements —such as Na, Cl, Mn, Al, Ti, and O. For much reactor NAA work, these container impurity levels are sufficiently low, relative to the levels of these and other elements in the samples, that they are of negligible importance. In such cases, the activated samples can be counted without transfer to fresh vials. For really low-level instrumental work, however, either the counting data must be corrected for the contributions from these container impurity elements (by calculation, or by spectrum subtraction), or the activated sample must be transferred to a fresh vial for counting. If samples are to be exposed to an nvt of greater than about $10^{17}/cm^2$, and/or to temperatures above about 80°C, other container materials—such as high-purity aluminium or quartz— should be used. In such cases, it is almost always necessary to transfer the activated sample to a fresh container for counting. Whereas the principal elements that constitute polyethylene (C and H) do not become significantly activated by reactor neutrons, Al, Si, and O are quite appreciably activated. Regardless of the type of container used, the samples are usually heat-sealed in their containers prior to activation—to prevent contamination of the reactor irradiation positions by spillage or leakage of sample. In thermal-neutron irradiations of samples with large overall neutron-capture cross sections, (n, γ) heating of the sample can be appreciable. Heating will be even more severe if the sample contains much lithium or boron, since they have large (n, α) cross sections, and the kinetic energy released is essentially entirely absorbed by the sample. Such high cross-section samples also produce considerable neutron self shielding. In such cases, one needs to limit the samples to very small ones, in order to minimize both heating and self shielding.

Certain kinds of sample materials, e.g. explosives, highly volatile materials, or explosively-polymerizable monomers cannot be safely activated unless very special provisions are made—such as use of high-pressure containers, vented containers, dilute solutions of the sample material, or very minute samples.

In general, samples exposed to reactor radiation are simultaneously exposed to appreciable fluxes of thermal neutrons, fast neutrons, and γ-ray photons. For example, in the 250 kW TRIGA Mark I reactor, in the rotary specimen rack, where the thermal-neutron flux is $1 \cdot 8 \times 10^{12}$ n cm^{-2} sec^{-1}, the flux of neutrons above 10 keV energy is $1 \cdot 5 \times 10^{12}$ n cm^{-2} sec^{-1}, and the γ-ray radiation level is 4,000 rads/sec. In the pneumatic tube position, the corresponding values are 4.3×10^{12}, 3.5×10^{12}, and 15,000 rads/sec. In long irradiations at high fluxes, certain kinds of samples—especially organic materials—can undergo appreciable radiation decomposition, caused by thermal-neutron capture recoils, fast-neutron collisions, and γ-ray radiolysis. It should be noted that all, or most, of the activity generated in a sample by (n, γ), (n, n'), or $(n, 2n)$ activation results in the radionuclides generated ending up in different chemical forms or valence states than the forms in which the elements are present in the original sample. This is because of the recoils suffered in the prompt decay of the compound nucleus when it emits one or more γ-ray photons, a neutron, or two neutrons, respectively. In (n, p) and (n, α) reactions, of course, the product radionuclide is a nuclide of a different element: one that is one unit lower in Z, in (n, p) reactions, or two units lower in Z, in (n, α) reactions, than the target nucleus.

It should also be remarked that, in typical reactor irradiations, only a minute fraction of the amount of an element present in a sample is converted to a different element. For example, in a 1 hr irradiation at a thermal-neutron flux of 10^{13} n cm^{-2} sec^{-1}, only 0.000004% of the amount of an element present, that has an (n, γ) cross section of 1 barn, undergoes reaction, i.e., 99.999996% of the original amount of element does not interact. The fractional conversion of an element may be calculated from the equation, $-\Delta N/N = \phi \sigma t_i$. In very long irradiations of elements with exceptionally large (n, γ) cross sections, at very high thermal-neutron fluxes, element consumption can, of course, become significant. For example, if an element contains a stable nuclide with an isotopic (n, γ) cross section of 10,000 barns, and the sample is exposed to a thermal-neutron flux of 10^{14} n cm^{-2} sec^{-1} for 100 hr, 30% of that nuclide would be changed to a nuclide of that element one mass number higher. If the product nuclide were radioactive, decaying by β^- emission, this would produce in the sample a corresponding amount of the stable nuclide of that mass number, but one unit higher in Z. For example, if a sample rich in dysprosium, but very low in holmium, were analyzed for holmium, a result somewhat in error would result if a correction were not made for the stable ^{165}Ho formed by the β^- decay of 2.33 hr ^{165}Dy, in turn formed from stable ^{164}Dy (28.18 atomic % of dysprosium) with a high thermal-neutron cross section (2,800 barns).

If the element of interest forms a neutron-interaction radionuclide of rather short half life (typically, in the range of seconds to minutes), only a short irradiation period is necessary, or even desirable. In the fundamental activation analysis equation, $A_o = N\phi\sigma S$, the saturation term, S, is $(1-e^{-0.693t_i/T})$, where t_i is the duration of the irradiation period and T is the half life of the radionuclide formed, both expressed in the same units of time. Thus, S, which is a dimensionless quantity that can only range between 0 and 1, acquires values of $\frac{1}{2}, \frac{3}{4}, \frac{7}{8}, \frac{15}{16}, \frac{31}{32}, \ldots$ at t_i/T values of 1, 2, 3, 4, 5. . . . It is, therefore, of little value to prolong the irradiation period to more than 2 or 3 times the half life of the radionuclide of interest. Longer irradiation increases the desired activity very little, and continues to build up the levels of interfering activities of longer half lives. One therefore employs short irradiations in such cases, using a pneumatic sample-transfer tube, irradiating samples one at a time—with rapid counting of each sample, before the activity of interest has decayed away. In the standard pneumatic tube used in the TRIGA Mark I reactor, for example, the transit time from the counting room into the reactor-core irradiation position, and the transit time back out to the counting room, are each about 2·5 sec. With the reactor running at the desired steady power level, the sample, in its polyethylene vial, in turn placed inside a more sturdy shock-resistant polyethylene rabbit, is sent into the core. The desired irradiation time is preset on an accurate timer. When this time has elapsed, the sample rabbit is automatically sent back out. As it comes out of the core position, it triggers two precise timers (via a photocell device). The rabbit is rapidly and automatically monitored, for safety, and then opened, the sample vial placed in the γ-ray spectrometer, and counting started. The start of the counting turns off one of the timers, thus giving an accurate value of the decay time from t_o to start of count. The sample is usually counted for a pre-selected live time (in the range of seconds to minutes, depending upon the half life of the radionuclide of interest). When the counting stops, the second timer stops, thus providing also an accurate value for the elapsed counting time (clock time). For low activity samples, of course, the percentage analyzer dead time will be very small, and the counting live time will be almost exactly the same as the counting clock time. In general, counting rates resulting in dead times greater than about 10% should be avoided, with NaI(Tl) detectors, as such high counting rates result in appreciable gain shifts. The clock counting time is used to make accurate decay corrections. To prevent contamination from Na and Cl from the hands, handling of the sample vial and rabbit is carried out with the use of clean plastic gloves—both before and after irradiation. Standard samples of the element of interest are interspersed in a series of unknown samples, and are activated and counted in exactly the same way as the unknown samples. Typical pneumatic tubes operate either with a partial vacuum on one side and atmospheric pressuron the other side, or by use of compressed air or nitrogen on one side and atmospheric pressure on the other side. Since the driving gas becomes appreciably activated, in the vicinity of the core (forming ^{16}N, ^{13}N, ^{15}O, ^{19}O, and ^{41}Ar), the system must be properly vented. In some kinds of research reactors, the pneumatic tube position is outside the reactor core, because of concern about abrupt reactivity changes that could occur if fissionable or high cross-section samples were rapidly injected into the core and/or rapidly ejected from the core. Because of the intrinsic safety conferred by the hydride type of fuel element used in the TRIGA type of reactor, the pneumatic tube terminus can be right in the core, in a vacant fuel element position.

When the element (or elements) of interest forms a longer-lived neutron activation product, i.e. one with a half life in the range of many minutes to hours, days, or longer, improved sensitivity is achieved by employing longer irradiation periods. Typically, irradiation periods of 30 min up to perhaps 10 hr may be employed—to increase the value of the saturation term in the activation equation. In such cases, it is more efficient to activate quite a number of samples, and standards, at the same time. Subsequently, they can be counted one at a time (or two or more at a time, if more than one γ-ray spectrometer is available), since the activity of interest is not decaying so rapidly. In such cases, there is also ample time to transfer each of the activated samples to a fresh container, if desired, prior to counting. Also, one usually allows some decay time, prior to start of the counting, to allow shorter-lived interfering activities to decay out. If radiochemical separations must be employed, this initial decay period is also helpful in minimizing the radiation exposure of the chemist.

In many kinds of research reactors, the various sample irradiation positions usually have somewhat different neutron flux values. Thus, if a whole group of samples are activated at one time, it is usually necessary—for work of the best accuracy—to attach a neutron flux monitor specimen (such as a tiny, weighed, piece of gold foil) to each sample vial. After irradiation, then, not only must the samples be counted, but also all of the monitor foils—the sample and standard results finally all being normalized to the same neutron flux. Alternately a flux-profile curve can be determined, using such monitor samples in all the different irradiation positions, and sample values subsequently normalized by use of this flux-profile curve. This is not such an accurate procedure, however, unless the curve is re-measured periodically—because of changes resulting from fission-product buildup and fuel consumption in the fuel elements with continued use of the reactor.

One type of research reactor that has an excellent provision for the simultaneous irradiation of many samples, at precisely the same average neutron flux, is the TRIGA reactor. As shown in the cutaway

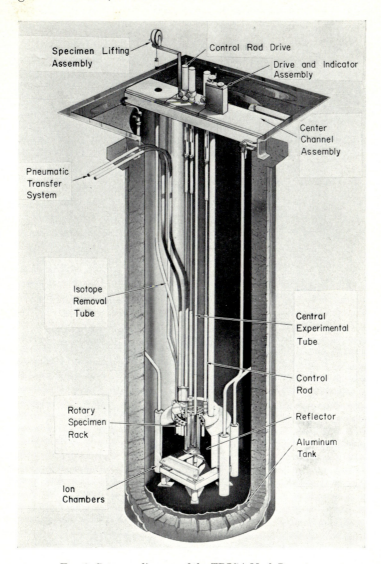

FIG. 2. Cutaway diagram of the TRIGA Mark I reactor.

diagram of Fig. 2, the core is surrounded by an annular ring of graphite, which serves as a neutron reflector and as a housing for an annular air cavity in which there is a rotary specimen rack, containing 40 tubes.

Samples, usually in half-dram, 2-dram, or 4-dram polyethylene vials, are placed in larger polystyrene tubes, each of which can be lowered into one of the 40 rotary rack tubes. When the rack has been loaded with as many samples as desired, the rack motor drive is turned on (1 r.p.m.), and the reactor is quickly brought up to power. At the end of the desired irradiation period, the reactor is shut down and rack rotation stopped. The samples are then removed, one tube at a time, each being automatically monitored, for safety, as it reaches the top of the pool. Provisions are made to handle any excessively radioactive samples with long-handled tools, and to store them temporarily in lead containers. Because of the rack rotation, each of the 40 irradiation positions receives exactly the same average neutron flux—thus greatly facilitating activation analysis work, since flux monitoring of each position is no longer necessary. Typically, if anywhere from 1 to 39 samples are to be analyzed for one particular element (one that forms a moderately long-lived radionuclide), one position in the rack is occupied by a standard sample of that element, of the same volume and shape as the unknown samples. Similarly, samples can be analyzed for two or more particular elements, along with separate standard samples of each, or a single standard sample that contains accurately known amounts of each. The polystyrene sample-vial containers can be lowered into the rack or withdrawn from it either by means of a manually-operated winch or fishing pole, or by means of a motor-driven device.

With many kinds of sample materials, samples as large as 10 grams can be employed. Where thermal-neutron self shielding is appreciable, however, sample sizes may have to be limited to amounts of the order of 1 g, 100 mg, 10 mg, or 1 mg—depending upon the overall thermal-neutron capture cross section of the sample material, per unit weight.

4. USE OF RECOIL PROTONS

When a sample contains a significant amount of hydrogen—such as aqueous solutions and most organic materials—energetic recoil protons are generated within the sample, by being struck with fast neutrons. In some cases, these recoil protons can induce nuclear reactions, such as (p, γ) or (p, n) reactions, in some elements contained in the sample. These can be of analytical use in some instances, or they can produce a small interference that must be corrected for. For example, studies in the author's laboratory and in some other laboratories have shown that samples of water can be analyzed instrumentally for their ^{18}O content, by measurement of the 110 min ^{18}F (a pure positron emitter) formed by the $^{18}O(p, n)^{18}F$ recoil-proton reaction. This reaction is endoergic, with a threshold value of 2·59 MeV. Even at a fission-spectrum neutron flux of only about $10^{12}\,n\,\text{cm}^{-2}\,\text{sec}^{-1}$, a 1 hr irradiation produces about 1,000 photopeak cpm of ^{18}F (0·511 MeV, 3×3 in NaI(Tl) detector) per gram of natural-abundance water, in which the ^{18}O is only 0·204 atomic % of the oxygen present. Under even these relatively low-flux conditions, the natural ^{18}O can still be barely detected by this method in as little as 10 mg of water. The technique has useful application in the field of ^{18}O tracer studies, where highly-enriched ^{18}O is introduced into a system under study, and resulting samples are then converted to water and analyzed for their ^{18}O contents. Any appreciable amount of fluorine must not be present in the samples, of course, as this would also form ^{18}F, via the ^{19}F$(n, 2n)^{18}$F fast-neutron reaction. This endoergic reaction has a threshold of 11·0 MeV, and an average fission-spectrum cross section of 0·05 millibarn.

An example in which a recoil-proton reaction can produce an undesirable interference is in the fast-neutron determination of nitrogen, via the ^{14}N$(n, 2n)^{13}$N reaction (10·5 MeV endoergic). If the sample contains appreciable amounts of carbon and hydrogen, some 9·96 min ^{13}N (a pure positron emitter) is formed by the ^{13}C$(p, n)^{13}$N recoil-proton reaction—the amount depending upon the quantities of H and C present in the sample, and upon the fast-neutron flux. This endoergic reaction has a threshold of 3·23 MeV. Even though ordinary carbon is only 1·107 atomic % ^{13}C, a typical pure hydrocarbon will produce enough ^{13}N, by this reaction, to appear to contain a few hundred ppm nitrogen. Oxygen can also produce some ^{13}N, via the ^{16}O$(p, \alpha)^{13}$N recoil-proton reaction.

Samples can also be analyzed for their oxygen (^{16}O) content, in the reactor, by means of a different charged-particle reaction, the ^{16}O$(t, n)^{18}$F reaction. The samples are each intimately mixed with an oxygen-free lithium compound, then exposed to a well-thermalized neutron flux for a period of the order of one hour. The high cross-section ^{6}Li$(n, \alpha)t$ reaction, which is exoergic to the extent of 4·79 MeV, produces energetic tritons, having initial energies of 2·73 MeV. These react with ^{16}O nuclei, via the (t, n) reaction, to form ^{18}F. This reaction is exoergic to the extent of 1·28 MeV (but there is a small Coulomb barrier, 0·63 MeV). This technique has been used to analyze samples for oxygen, but has a number of complications, and has largely been superseded by the 14 MeV neutron method, employing a small Cockcroft-Walton deuteron accelerator to produce 14 MeV neutrons (via the ^{3}H$(d, n)^{4}$He reaction), the ^{16}O$(n, p)^{16}$N reaction, and detection of the 6·13 MeV and 7·11 MeV γ-rays of 7·14 sec ^{16}N. The ^{16}O$(n, p)^{16}$N reaction is endoergic, with a threshold of 10·2 MeV.

5. USE OF DELAYED NEUTRONS

If a sample containing ^{235}U is activated briefly (perhaps for one minute) in a reasonably high thermal-neutron flux, and then promptly counted in a neutron detector, it is possible to count the delayed neutrons resulting from the neutron-emission decay of some of the fission products produced by the fission of ^{235}U—and to use this as a sensitive means of determining ^{235}U purely instrumentally. If the artificially-produced nuclides, ^{233}U and/or ^{239}Pu, should also be present, they would interfere, since they are also fissioned by thermal neutrons, also generating delayed neutrons. If the sample is surrounded by cadmium or boron during irradiation, and is exposed to a high fast-neutron flux, the same technique can be used to determine ^{238}U and ^{232}Th. These nuclides are not fissioned by thermal neutrons, but do undergo fission when interacting with neutrons having energies above their (n,f) threshold—about 1 MeV. The usual counter employed for such work is an array of several BF$_3$ (or ^{3}He) thermal-neutron detectors, imbedded in a tank of paraffin, the counters arranged around the activated sample.

6. ANALYSIS VIA PROMPT γ-RAYS

If a beam of predominantly thermal, or predominantly fast, neutrons is used (originating in or near the reactor core), it is possible to utilize the prompt thermal-neutron capture γ-rays, or the prompt neutron inelastic scattering γ-rays for the detection of some elements present in a sample exposed to such a beam, instead of the radioactive-decay γ-rays. If an essentially thermal-neutron beam is desired, the beam tube should include an appreciable section of either graphite or D$_2$O (for neutron thermalization with minimum loss by capture), usually followed by a section of bismuth—which serves to absorb the γ-ray component of the beam without excessive capture of neutrons. If a predominantly fast-neutron beam is desired, the beam tube should include a cadmium filter and a bismuth section. If the sample is viewed by a γ-ray spectrometer, well-shielded not only by lead, but also by lithium or boron (in a moderator matrix, and placed outside the lead), the prompt γ-rays emitted by the sample can be detected while the sample is exposed to the neutron beam. The detector is arranged so as to view the sample at right angles to the direction of the neutron beam, through a neutron-absorbing, but γ-ray transmitting, window in the detector shield. Lithium-6 and boron-10 are both good thermal-neutron absorbers for use in the shield and window, in that they have very high thermal-neutron absorption cross sections, and their interaction is via an (n, α) reaction, rather than via an (n, γ) reaction.

Because of the fact that the neutron fluxes available from such beam tubes are much lower than those available in or near the reactor core (perhaps 10^6–10^7 n cm^{-2} sec^{-1}, instead of 10^{12}–10^{13} n cm^{-2} sec^{-1}), the fact that half life is no longer a useful variable, and the fact that the prompt-gamma spectra of most elements are quite complicated, prompt-gamma neutron activation analysis is rather limited in its applications. For many matrices, however, it can be used effectively to determine the major and minor (rather than trace) elements in the samples. The prompt γ-rays emitted by many elements range all the way from less than 0.1 MeV up to even 10 MeV. One feature of this technique is that the production of γ-rays by a particular nuclide, upon interaction with neutrons, depends only upon the interaction cross section—not upon the half life of the product (if it is a radionuclide) or even upon whether the product nuclide is radioactive. Thus, for example, the thermal-neutron ^{155}Gd(n, γ) ^{156}Gd reaction, which has an isotopic cross section of 56,200 barns, generates capture γ-rays, even though ^{156}Gd is a stable nuclide of gadolinium—not a radionuclide. The capture γ-rays are emitted, of course, by the prompt ($\sim 10^{-13}$ sec) decay, in this example, of the [^{156}Gd]* excited compound nucleus. Hydrogen, which is not detected by ordinary NAA, can be determined via the capture γ-rays resulting from the ^{1}H$(n, \gamma)^2$H thermal-neutron reaction.

Similarly, many stable nuclides can be detected via the prompt neutron-inelastic-scattering γ-rays produced as a result of an (n, n') reaction—even though the product is not a metastable isomer of the original nucleus. Thus, carbon, which is not appreciably activated via thermal- or fast-neutrons (at least up to quite high energies), to form a radionuclide that can be subsequently detected, can be detected via the prompt γ-rays resulting from the ^{12}C$(n, n')^{12}$C fast-neutron interaction.

Similar thermal-neutron reactor beam tubes are also of use for several kinds of non-activation analysis applications, such as the neutron-absorption determination of high cross-section elements such as Li, B, Cd, etc.; neutron radiography, and neutron diffraction. Recently, following a suggestion by J. M. A. Lenihan, they have also been used for *in vivo* activation analysis measurements—such as determining the iodine content of the thyroid gland in animals and in man.

G. Use of High-Intensity Reactor Pulses

For reasons of safety, most types of nuclear reactors are only operated at a steady, closely-regulated, power and flux level, once they have been brought up to the desired power level. Because it uses a hydrided zirconium-uranium type of fuel element, however, the TRIGA type of research reactor can also be pulsed, safely, frequently, and reproducibly, to extremely high power and flux levels, when desired. This is possible because of an interesting intrinsic phenomenon. When a pulse is desired, a prescribed amount of excess reactivity (up to about 3·5% $\delta k/k$) is abruptly introduced, via pneumatic ejection of a control rod. Prior to this, the reactor is operated steadily at a very low power level (typically, 5 W). The core thus goes rapidly from a just-critical condition to an appreciably supercritical condition. The fission reaction rapidly rises, exponentially, reaching a power level, within a few milliseconds, of the order of 10^9 W. Since the hydrogen atoms in the fuel elements promptly rise in temperature, with no delay, as the fission rate rises, they reach temperatures of several hundred °C within this same short time period. Since hydrogen nuclei at such temperatures can only, at best, moderate fission neutrons down to an energy corresponding to kT, there is a prompt shift in the low-energy end of the neutron spectrum. Since the fission cross section of ^{235}U decreases sharply with increasing neutron energy (decreasing, for example, from 582 barns at 0·025 eV to 380 barns at 0·050 eV), the system suddenly becomes subcritical—as a result of the temperature increase within the fuel elements—and the reactor shuts itself off.

In the author's laboratory, the 250 kW (steady power) TRIGA Mark I reactor is regularly pulsed to peak powers as high as 1,000 MW (2·2% excess reactivity), and the 1·5 MW (steady power) Advanced TRIGA Prototype reactor (ATPR) is regularly pulsed to peak powers as high as 6,400 MW (3·2% excess reactivity). The Mark I reactor utilizes aluminium-clad fuel elements with a 1/1 H/Zr atomic ratio; whereas the ATPR employs stainless steel-clad elements with a 1·7/1 H/Zr ratio, an additional ring of fuel elements (the G ring), and water (rather than graphite) as reflector. In a 1,000 MW Mark I pulse, the peak total neutron flux, in the pneumatic tube position, is about 3×10^{16} n cm^{-2} sec^{-1}, and the pulse duration (full width at half maximum) is about 14 msec. In a 6,400 MW ATPR pulse, the corresponding figures are about 2×10^{17} n cm^{-2} sec^{-1} and about 6 msec. The Mark I reactor can be reproducibly pulsed to 1,000 MW up to 10 times per hour; the ATPR, to 6,400 MW, up to six times per hour. A recorder chart tracing of a 900 MW Mark I pulse is shown in Fig. 3.

Such high-intensity pulses have been shown to be of value in NAA work, when the desired induced activity is very short-lived. In experimental studies in the author's laboratory (Yule and Guinn, 1964), it was shown that, with 1,000 MW pulses of the TRIGA Mark 1 reactor,

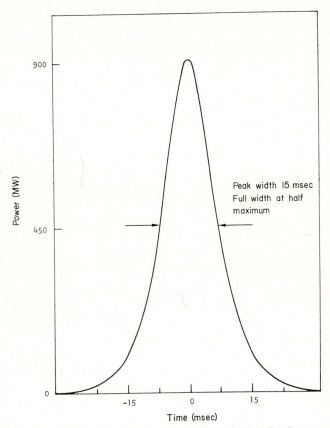

FIG. 3. Record of a 900 MW TRIGA Mark I pulse.

the ratio of the activity generated in a pulse, A_p (of a particular radionuclide) to the activity level of that radionuclide that could be produced, even at saturation, by operation of the reactor at the normal steady-power level of 250 kW, A_s, was equal to $70/T$, where T is the half life of the induced activity, expressed in seconds:

$$A_p/A_s = 70/T. \tag{5}$$

In other words, under these conditions, a pulse provides about a 700-fold improvement in sensitivity for an element that forms a product with a half life of 0·1 sec, 70-fold improvement in sensitivity for an element that forms a product with a half life of 1 sec, 7-fold if the half life is 10 sec, and no improvement if the half life is 70 sec. For longer-lived induced activities, a longer irradiation at the steady power level of 250 kW gives a better sensitivity than is possible with a pulse.

The general relationship is readily derived, as follows. In a pulse that is very short in duration, compared with the half life of the product formed,

$$A_p = N\sigma(\text{nvt})\, 0.693/T, \tag{6}$$

where nvt is the total thermal (or fast) neutron dose, per cm², received by the sample during the pulse, and σ is the thermal- (or fast-) neutron reaction cross section. With a 1,000 MW pulse (peak power) with a 14 msec half width, the integrated energy is 15 MWs (i.e. 14×1.064). This corresponds to a thermal-neutron nvt, in the pneumatic-tube position, of 2.6×10^{14} n/cm².

In a steady-state irradiation at 250 kW, the thermal-neutron flux in this same position is $4.3 \times 10^{12}\, n$ cm^{-2} sec^{-1}. In a steady irradiation to saturation,

$$A_s = N\phi\sigma. \tag{7}$$

The ratio of A_p to A_s is thus $0.693\,(\text{nvt})/\phi T$, or $0.693 \times 2.6 \times 10^{14}/ 4.3 \times 10^{12}\,T$, or $42/T$. In one main series of pulsing experiments carried out, the integrated pulse energies were 19.5 ± 0.5 MWs, in each case, instead of 15 MWs, thus resulting in an expected relationship of $A_p/A_s = 54/T$. Experimentally, for some 15 induced activities studied —some thermal-neutron products, some fast-neutron products, the experimentally-observed relationship was that shown in Equation (5), i.e. $A_p/A_s = 70/T$. The ratio of A_p/A_s is of course the same, for a given radionuclide, at any given decay time, although A_p and A_s in Equations (6) and (7) represent the values at zero decay time. Since thermalization of fission neutrons is very rapid, even compared with the pulse duration, the neutron-energy spectrum during a pulse should be essentially the same as that during steady operation—hence the same relationship should apply to both thermal-neutron products and fast-neutron products. Experimentally, such proves to be the case. Three pure (n, γ) products (11.6 sec ^{20}F, from F, 19.5 sec ^{46m}Sc from Sc, and 14 sec ^{199m}Pt from Pt) gave A_p/A_s ratios of $56/T$, $70/T$, and $71/T$, respectively. Two other activities that are largely generated by (n, γ) reaction, but also with some fast-neutron contribution from concurrent (n, n') and $(n, 2n)$ reactions, 17.5 sec ^{77m}Se from Se, and 5.3 sec ^{183m}W from W, gave A_p/A_s ratios of $112/T$ and $41/T$, respectively. The remaining 11 reactions studied were reactions produced only by fast neutrons:

7·14 sec ^{16}N from ^{16}O via (n, p)	73/T
29·1 sec ^{19}O from ^{19}F via (n, p)	82/T
37·6 sec ^{23}Ne from ^{23}Na via (n, p)	88/T
11·6 sec ^{20}F from ^{23}Na via (n, α)	54/T
60 sec ^{25}Na from ^{25}Mg via (n, p)	162/T
12·4 sec ^{34}P from ^{34}S via (n, p)	95/T
225 sec ^{52}V from ^{52}Cr via (n, p)	56/T
48 sec ^{75m}Ge from ^{75}As via (n, p)	62/T
16·1 sec ^{89m}Y from ^{89}Y via (n, n')	31/T
10·5 sec ^{158m}Tb from ^{159}Tb via $(n, 2n)$	97/T
7·2 sec ^{197m}Au from ^{197}Au via (n, n')	66/T
Mean value:	79/T

If the two rather extreme values (162/T for ^{25}Na, and 31/T for ^{89m}Y) are rejected, the mean value for these fast-neutron products becomes 74/T. Considering the rather large (and as yet unexplained) variation in the experimental A_p/A_s ratios, from one reaction product to another, the fast-neutron product relationship of 74/T is in good agreement with the mean value from the five (n, γ) or predominantly (n, γ) products 70/T. A least-squares fit of all the data gives a value of $(70 \pm 20)/T$.

Such pulsing can be done very reproducibly. For example, in two series of pulses (four pulses in each series) that were nominally supposed to be 900 MW pulses, the measured peak powers ranged only from 880 MW to 910 MW, with a standard deviation from the mean value of 896 MW of only $\pm 1·3\%$. Similarly, the reproducibility of generation of a particular induced activity is very good. For example, in five consecutive nominally 1,000 MW pulses, the yield of ^{46m}Sc from one particular sample gave values of $(2·57 \pm 0·11) \times 10^{12}$ photopeak cpm/g Sc, i.e. the standard deviation from the mean value was only $\pm 4·3\%$ of the value.

More recently, additional short-lived species have been studied, in the author's laboratory, by H. R. Lukens. Of particular interest are the radionuclides, 0·80 sec ^{207m}Pb (from lead) and 0·84-sec ^{8}Li (from lithium, and from boron). The ^{207m}Pb emits γ-rays of 0·570 MeV and 1·064 MeV energies, and a 1,000 MW pulse gives an 88-fold improvement in sensitivity, compared with a normal 250 kW irradiation in the same position, to saturation. With such a pulse, the limit of detection for lead is about 0·3 μg. The ^{207m}Pb is formed concurrently by the ^{206}Pb$(n, \gamma)^{207m}$Pb, ^{207}Pb$(n, n')^{207m}$Pb, and ^{208}Pb$(n, 2n)^{207m}$Pb reactions. Lithium-8 is formed from lithium by the thermal-neutron ^{7}Li$(n, \gamma)^{8}$Li reaction, and from boron by the fast-neutron $^{11}B(n, \alpha)^8Li$ reaction. The $^7Li(n, \gamma)^8Li$ thermal-neutron reaction has an isotopic cross section of only 0·036 barn, yet, with the 83-fold improvement in sensitivity provided by a 1,000 MW pulse, as little as 0·0006 μg of lithium can be detected. The sensitivity for boron, using such a pulse, is not as good, namely, 1 μg, but is still quite useful. Whether the ^8Li is being formed from lithium or boron, in an unknown sample, can be determined by measurements with, and without, cadmium around the sample during activation. It is interesting to note that with both of these elements, it is the low (n, γ) cross-section, higher-abundance, stable nuclide (^7Li and ^{11}B) of the element that is thus determined—not the high (n, α) cross-section, lower-abundance nuclide (^6Li and ^{10}B). Since ^8Li is a pure β^- emitter (13·1 MeV E_{max}), it is detected by means of a well-type Cerenkov counter, equipped with a liner that absorbs β^- particles of energies below about 4 MeV, and multichannel scaling.

For studies with very short-lived species, especially those with half lives in the range of a few tenths of a second to a few seconds, a separate, high-speed, pneumatic tube is used. The sample is placed in a specially-constructed sturdy-polyethylene vial (with no secondary container), then sent into the irradiation position (another vacant fuel-element position in the F-ring of the TRIGA Mark I reactor), the reactor brought up to a few watts of power, and then pulsed. The completion of the pulse triggers the ejection of the sample directly out into the γ-ray spectrometer (or Cerenkov counter), and initiates counting. Whereas the usual pneumatic tube (2·5 sec transit time) functions with a vacuum system and normal air pressure, this faster one employs compressed nitrogen gas. Using driving pressures as high as 80 psig, sample transit times as short as 0·2 sec can be achieved.

To date, about 20 different short-lived neutron-induced activities have been studied, in the author's laboratory, by means of high-intensity reactor pulses. Additional ones remain to be investigated, and further studies along these lines are in progress.

H. New Techniques and New Areas of Application

Some relatively new techniques that are already considerably extending the capabilities of reactor neutron activation analysis should be mentioned briefly. In particular, the purely-instrumental form of the method is being greatly advanced by the use of lithium-drifted germanium semiconductor detectors, and by the use of advanced computer techniques. The Ge(Li) detector typically has an energy resolution some 15 times better than that of a NaI(Tl) scintillation detector.

Whereas, for example, the latter has about a 5% resolution at a γ-ray energy of 1 MeV (i.e. 50 keV), the Ge(Li) detector resolution at this energy is about 0·3% (i.e. 3 keV). As larger Ge(Li) detectors become available, they will become increasingly useful in NAA γ-ray spectrometry. The largest ones currently available are only about 40 cm^3 in volume, with a maximum sensitive depth of about 2 cm. They must be operated and maintained at liquid-nitrogen temperature. For maximum resolution, they are operated with a liquid-nitrogen cooled field-effect transistor (FET) preamplifier, a high-quality pulse amplifier, and a 4096-channel pulse-height analyzer. Because of the large number of data points from even one spectrum, a computer-compatible magnetic-tape readout is desirable, followed by computer data reduction.

Both with NaI(Tl) and Ge(Li) pulse-height spectra, advanced computer techniques for exhaustive processing of the pulse-height data are becoming increasingly used. Computer programs are available to (1) correct the data for any gain shift, (2) smooth the data by a convolution process, (3) search for photopeaks and calculate their energies, (4) compute net photopeak areas and compare with standards (making all necessary decay and flux corrections) to give μg and ppm values for each element detected, each with its standard deviation, (5) carry out a weighted least-squares fit of the data, with resolution into the various radionuclide contributions, and (6) carry out a resolution of the contributions to a given photopeak from two or more radionuclides of essentially the same γ-ray energy, but of different half lives (especially important with β^+ emitters). Such computer processing of the pulse-height data is extremely valuable when complicated NaI(Tl) spectra, perhaps including a number of overlapping photopeaks, need to be quantitatively resolved into all the statistically-significant components. Although the use of Ge(Li) usually eliminates the problem of overlapping photopeaks, the much larger number of data points makes computer processing of the data—as opposed to manual calculation—highly attractive.

In cases where the purely-instrumental technique is not adequate, due to serious interfering activities, radiochemical separations with carriers and hold-back carriers are usually employed. Traditionally, this has been a very tedious process, especially if one is endeavouring to determine many elements in a sample, rather than just one or a few. A new advance that holds great promise in this area is the development of a commercially-available automated radiochemical-separation device, in Sweden. Following introduction of an aqueous sample into the unit (such as wet-ashed activated biological sample), separation of 48 elements into 13 groups. on 13 small ion-exchange columns, is achieved within 40 mins (Erwall, 1967). Each column can then be counted directly on a γ-ray spectrometer.

Because of the very great sensitivities attainable for a large number of elements by means of high-flux NAA, the method has found—and is increasingly finding—many valuable applications in almost every branch of science, industry, medicine, and agriculture. Important applications are being made in such specialized fields as archaeology, oceanography, space exploration, and criminalistics. In the biomedical field, important advances have been made in our knowledge of trace elements, and the method has been used productively in combination with the use of enriched stable-isotope tracers (^{58}Fe, ^{50}Cr, and ^{46}Ca). With the advent of some of the newer techniques described earlier in this chapter, applications of high-flux neutron activation analysis should advance at an even greater pace.

References

Buchanan, J. D. (1961). *In* "Modern Trends in Activation Analysis", College Station, Texas, p. 72; also General Atomic Report GA-2662.
Erwall, L. G. (1967). *Private Communication.*
Heath, R. L. (1964). "Scintillation Spectrometry Gamma-Ray Spectrum Catalogue," 2nd Edition (IDO–16880–1 and –2).
Hughes, D. J. and Schwartz, R. B. (1958). "Neutron Cross Sections," 2nd Edition (BNL–325).
Lukens, H. R. (1964). "Estimated Photopeak Specific Activities in Reactor Irradiations" (General Atomic Report GA–5073).
Roy, J. C. and Hawton, J. J. (1960). "Table of Estimated Cross Sections for (n, p), (n, α), and $(n, 2n)$ Reactions in a Fission Neutron Spectrum" (AECL–1181).
Strominger, D., Hollander, J. M. and Seaborg, G. T. (1958). *Rev. mod. Phys.*, **30**, 585.
Watt, B. E. (1952). *Phys. Rev.*, **87**, 1937.
Yule, H. P. and Guinn, V. P. (1964). *In* "Radiochemical Methods of Analysis", Vol. 2. p. 111. International Atomic Energy Agency, Vienna, 1965.

Bibliography

Ashby, V. J. and Catron, H. C. (1959). "Tables of Nuclear Reaction Q Values." *UCRL*–5419.
Atomic Energy Commission, U.S. (1955). "Research Reactors." McGraw-Hill, New York.
Besancon, R. M. (Ed.) (1966). "The Encyclopedia of Physics." Reinhold, N.Y.
Bock-Werthmann, W. (1961, 1963, 1964). "Activation Analysis Bibliography." *AED–C*–14–01, –02, –03. Max-Planck Institute, Frankfurt.
Bowen, H. J. M. and Gibbons, D. (1963). "Radioactivation Analysis." Clarendon Press, Oxford.

Camp, D. C. (1967). "Applications and Optimization of the Lithium-Drifted Germanium Detector System." *UCRL*–50156.
Comar, D. (Ed.) (1964). "L'Analyse par Radioactivation et Ses Applications aux Sciences Biologiques." Presses Universitaires de France, Paris.
Evans, R. D. (1955). "The Atomic Nucleus." McGraw-Hill, New York.
Friedlander, G., Kennedy, J. W. and Miller, J. M. (1964). "Nuclear and Radiochemistry," 2nd Edition. Wiley, New York.
General Atomic (1967). "Symposium on Research Reactor Applications, Abstracts."
Georgia Institute of Technology (1963). "Symposium on the Utilization of Research Reactors." *ORO*–3191–1.
Girardi, F., Guzzi, G. and Pauly, J. (1965). "Data Handbook for Sensitivity Calculations in Neutron Activation Analysis." *EUR*–1898e.
Glasstone, S. (1955). "Principles of Nuclear Reactor Engineering." Van Nostrand, New York.
Glasstone, S. and Edlund, M. C. (1952). "The Elements of Nuclear Reactor Theory." Van Nostrand, New York.
Goldman, D. T. and Stehn, J. R. (1961). "Chart of the Nuclides." General Electric Company, Schenectady, New York.
Greenwood, R. C. and Reed, J. H. (1965). "Prompt Gamma Rays from Radiative Capture of Thermal Neutrons." *IITRI*–1193–53, Vol. 1 and 2.
Guinn, V. P. (Ed.) (1967). "Proceedings of the First International Conference on Forensic Activation Analysis." General Atomic Report GA–8171.
International Atomic Energy Agency (1961). "Programming and Utilization of Research Reactors." Academic Press, London.
International Atomic Energy Agency (1962a). "Utilization of Research Reactors." *IAEA*, Vienna.
International Atomic Energy Agency (1962b). "Production and Use of Short-Lived Radioisotopes from Reactors." 2 vols., *IAEA*, Vienna.
International Atomic Energy Agency (1964). "Radiochemical Methods of Analysis." 2 vols., *IAEA*, Vienna.
International Atomic Energy Agency (1967). "Nuclear Activation Techniques in the Life Sciences." *IAEA*, Vienna.
Koch, R. C. (1960). "Activation Analysis Handbook." Academic Press, New York.
Lenihan, J. M. A. and Thomson, S. J. (Eds.) (1965). "Activation Analysis—Principles and Applications." Academic Press, New York.
Liskien, H. and Paulsen, A. (1963). "Compilation of Cross Sections for Some Neutron-Induced Threshold Reactions." *EUR*–119e.
Lukens, H. R. (1964). "Elemental Survey Analysis by Neutron Activation: Simplified Estimation of Upper Limits." General Atomic Report GA–5896.
Lukens, H. R. (1966). *Analytica chin. Acta* **34**, 9.
Lyon, W. S., Jr. (Ed.) (1964). "Guide to Activation Analysis." Van Nostrand, New York.
Meinke, W. W. and Scribner, B. F. (Eds.) (1967). "Trace Characterization—Chemical and Physical." National Bureau of Standards Monograph 100.
Mikesell, R. E. (1966). "TRIGA Experimental and Irradiation Facilities for Research and Development." General Atomic Report GA–1695, Rev. 5.
Morrison, G. H. (Ed.) (1965). "Trace Analysis: Physical Methods." Interscience, New York.

Samsahl, K., Wester, P.O. and Landström, O. (1968). " An Automatic Group Separation System for the Simultaneous Determination of a Great Number of Elements in Biological Material." *Anal. Chem.*, **40**, 181.

Sederer, C. M., Hollander, J. M. and Perlman, J. (1967). " Table of Isotopes. Sixth Edition." Wiley, New York.

Sutz, G. J., Borani, R. J., Maddock, R. S. and Meinke, W. W. (1968). " Activation Analysis: A Bibliography." NBS Technical Note 467, Parts 1 and 2. Supt. of Documents, U.S. Govt. Printing Office, Wash. D.C.

Taylor, D. (1964). "Neutron Irradiation and Activation Analysis." Van Nostrand, New York.

Texas A & M University (1961). "Modern Trends in Activation Analysis." College Station, Texas.

Texas A & M University (1965). "Modern Trends in Activation Analysis." College Station, Texas.

Wing, J. and Wahlgren, M.A. (1965). "Detection Sensitivities in Thermal-Neutron Activation." *ANS*–6953.

Yule, H. P. (1964). "Experimental Reactor Thermal-Neutron Activation Analysis Sensitivities." General Atomic Report GA–5631, and *Anal. Chem.* (1965), **37**, 129.

Yule, H. P., Lukens, H. R. and Guinn, V. P. (1964 and 1965). "Utilization of Reactor Fast Neutrons for Activation Analysis." General Atomic Reports GA–5073 and GA–5691, and *nucl. Instrs. & methods*, (1965), **33**, 277.

AUTOMATION AND ELECTRONIC DATA HANDLING

R. E. WAINERDI

Activation Analysis Research Laboratory,
Department of Chemical Engineering,
Texas A & M University,
College Station, Texas, U.S.A.

A. Introduction to Instrumental Activation Analysis and Automated Systems	81
1. Scope of instrumental methods and systems	81
2. General concepts of instrumental spectrometry	85
3. Activators and their integration into automated systems	86
4. The detector, analyzer, read-out complex and possible variations commonly encountered	89
5. Packaging of samples	89
6. Sample storage and retrieval	91
7. Noise control	92
8. Drift control	93
B. Data Production Considerations	94
1. Read-out format design	94
2. The types of read-outs available	95
3. Data handling prior to computer manipulation	95
C. Electronic Date Handling and Processing	96
1. Spectrum resolution considerations	96
2. The law of linear superposition	97
3. Spectrum stripping methods	97
4. Matrix methods	98
5. Least squares curve fitting	98
D. Modern Trends	99
References	99

A. Introduction to Instrumental Activation Analysis and Automated Systems

1. SCOPE OF INSTRUMENTAL METHODS AND SYSTEMS

Nuclear activation analysis was first proposed by George Charles de Hevesy and Hilda Levi in 1936, and until 1955 most activation analyses were performed via radiochemical separations, by first bombarding a sample in a flux of activating particles (usually neutrons), and then separating and purifying the various activated isotopes followed by a counting of the purified radioactivities.

Radiochemical activation analysis is extremely accurate and sensitive, and it avoids carrier contamination entirely. It is often the procedure of choice when extreme sensitivity is necessary. Radiochemistry is time consuming in those cases where large numbers of separations are necessary, such as when a sample contains a number of different constituents of interest, but excellent methods have been developed and are available for most separations.

The development of the modern multi-channel analyzer in the mid 1950's by Schumann and McMahon (1956), made it possible to measure the complete composite γ-ray spectrum, as emitted by an activated sample, and as detected by a sodium iodide crystal. It was soon demonstrated that γ-ray spectra could be unsorted into constituents through the use of various mathematical procedures, such as the method of Covell (1959), thus revealing the extent of the constituent γ-ray activities and, therefore, the relative mass of each elemental constituent.

During the period 1936–1956, most activation analyses were performed radiochemically, but only a small fraction of analyses are now performed in this way, although there is a growing interest in methods which combine simplified radiochemical separations with instrumental activation analyses.

Early work in the field of instrumental activation analysis indicated potentials for development both in automation of sample handling systems and in electronic data processing to cope with the large quantities of data which could be produced by automated systems. The early efforts of Salmon (1961), Kuykendall and Wainerdi (1960), and others, resulted in the development of many computer programs which perform spectrum unsorting using the high speed digital computer.

Computers were used in 1959 and 1960, for the unsorting of γ-ray spectra consisting of about five components. Work was initiated at that time by Fite, Kuykendall, and Wainerdi to build automated activation analysis systems which would be capable of irradiating and counting samples, and also preparing their spectral signature for computer processing. Figure 1 shows the first such system, called the Mark I. The basic operating parts and principals of this system are generally applicable using any activator, detector, data analyzer, and computer which may be available.

It is desirable to optimize the parameters of the analysis, such as irradiation, waiting, counting times, flux and geometry, but once these factors are optimized, similar samples can be processed automatically

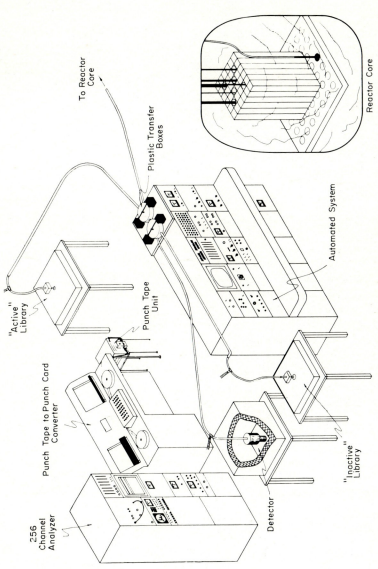

Fig. 1. Mark I-Ia automatic activation analysis system.

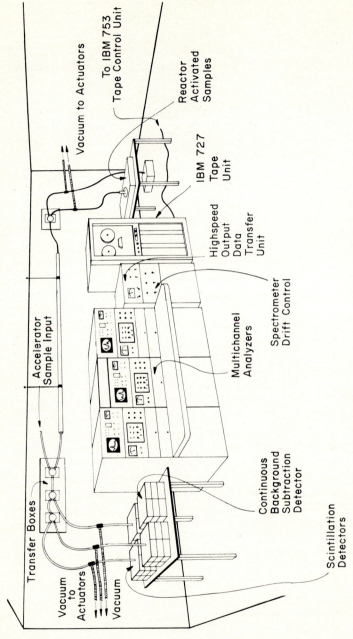

FIG. 2. Mark II automatic activation analysis system.

at fairly rapid rates. A more modern system (Fig. 2), involves three detectors and three analyzers coupled to a magnetic tape unit, and is called the Mark II System.

2. GENERAL CONCEPTS OF INSTRUMENTAL SPECTROMETRY

It is important, in discussing instrumental γ-ray spectrometry, to recognize that the principal competitive advantage of radiochemical separations lies in both the avoidance of carrier contamination and the

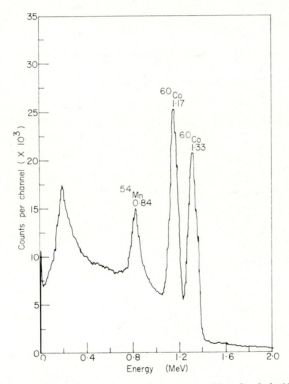

FIG. 3. Composite spectrum of manganese-54 and cobalt-60.

excellent specificity of the various separation procedures. These very important advantages in quantitative analysis can be lost instrumentally by any electronic or physical manipulation which interferes with the quantitative values of the pulse height distributions in the spectrum. In order to illustrate this point, observe Fig. 3, a spectrum consisting of cobalt and manganese γ-ray activities. In this spectrum, the cobalt-60 γ-ray peaks at 1·17 and 1·33 MeV consist entirely of ^{60}C activity but the manganese peak at 0·84 MeV contains Compton scattered γ-rays (from ^{60}Co) as well as the ^{54}Mg γ-rays. Thus the total area of the manganese peak includes components attributable to manganese and cobalt and, of course, background.

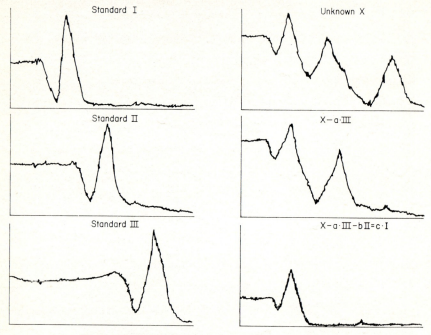

Fig. 4. Illustration of the "law of linear superposition".

The "Law of Linear Superposition" applies to γ-ray spectra collected under carefully controlled conditions, and the composite spectra, will be a linear addition of the individual spectra, as illustrated in Fig. 4, only when these conditions are observed. Typical factors which influence superposition are drift, high dead time, and electronic noise.

3. ACTIVATORS AND THEIR INTEGRATION INTO AUTOMATED SYSTEMS

Various types of nuclear activators have been utilized in activation analyses. Among the most prominent are the nuclear reactor, the 14 MeV neutron generator, the cyclotron, the linear accelerator, the Van de Graaff accelerator and the isotopic neutron source. Each of these devices produces one or more nuclear particles which possess the property of inducing characteristic radioactivity in an irradiated

Fig. 5. Triga reactor core.

sample. Samples and standards are irradiated under conditions which make it possible to vary times and other irradiation parameters. The integration of an activator into an automated activation analysis system involves either having the activator "on" continuously, or else triggering the activator "on" as the sample approaches the irradiation position. An example of this type of situation involves the pulsing of a nuclear reactor, Fig. 5. A sample passes a photocell which triggers a

FIG. 6. Illustration of sample transfer box.

safe reactor pulse of many millions of watts, and during the instant of maximum power, the sample passes closest to the core, and is counted a few milliseconds later. In order for this procedure to be useful in activation analysis, it is vital that the time between the triggering of the pulse and the pulse be very carefully controlled so that the sample is in the proper activation position to receive the neutron dose at the peak of the reactor pulse. More commonly, nuclear reactors or other activating sources are run in a steady state condition and pneumatic tubes introduce samples and withdraw them automatically at pre-set times Transfer boxes such as those shown in Fig. 6 can be built out of plastic, without lubricants, and can be used for activation analysis in such a way as to preclude sample contamination. By use of such transfer boxes, samples can be directed throughout a complex system such as that shown in Fig. 1 in which the activator, in this case a nuclear reactor is connected with a detector, multi-channel analyzer, and read-out complex.

4. THE DETECTOR, ANALYZER, READ-OUT COMPLEX AND POSSIBLE VARIATIONS COMMONLY ENCOUNTERED

Automatic, computer-coupled activation analysis was developed at Texas A & M University in 1959–61 by Wainerdi, Gibbons *et al.* (1962). The first such system consists of a completely integrated system presenting data at the output in computer format. Figure 2 shows the Mark II automatic activation analysis system, a later development (1961–63) which consists of three analyzers and a read-out, all interconnected with pneumatic tubes, and controlled in such a fashion that the system can either accept samples from bulk irradiation in a reactor or cyclotron, and in the reactor case, samples activated either singly or in groups.

Another type of automated activation analysis system is the Mark III system built in 1961–63 and shown in Fig. 7. This system consists of two detectors connected to a single multi-channel analyzer, a pneumatic tube and a 14 MeV neutron generator. It is probably the most widely duplicated activation analysis system existing in virtually identical form in at least 15 countries. There are a number of other detector, analyzer, read-out configurations which are possible, and which include such variations as the inclusion of a simple radiochemistry step between the activation and the counting stages; this permits one or two gross contaminants or interferences to be removed to simplify later γ-ray spectrometry. For example, the removal of radiosodium by precipitation (Menon and Wainerdi, 1965), with HCl and n-butanol, can be performed at this stage.

5. PACKAGING OF SAMPLES

In an automated system, where the irradiated sample is counted in its original container, it is important to minimize the induced activity which might be contributed by the material of the container or by the argon in the contained air. Polyethylene vials such as those shown in Fig. 8 are commonly used for this purpose. Aluminium or quartz containers are sometimes used where very high neutron fluxes, high temperatures or high pressures are encountered. In the case of quartz

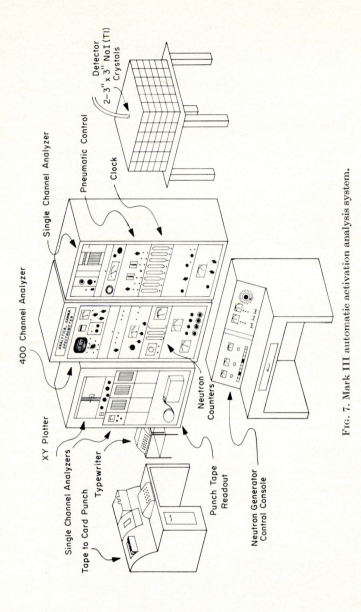

Fig. 7. Mark III automatic activation analysis system.

and aluminium containers, it is occasionally necessary to remove the sample from its irradiation container and place it in a second container for instrumental γ-ray spectrometry because of the many impurities common in quartz and aluminium. Such transfer is a difficult step to automate, and is one that is commonly done manually in most laboratories.

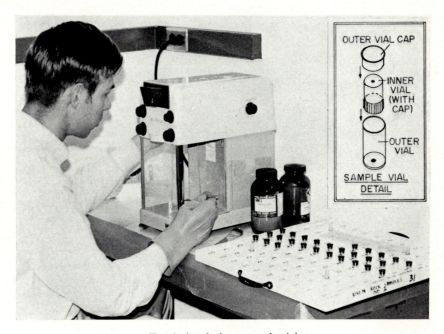

FIG. 8. A polythene sample vial.

6. SAMPLE STORAGE AND RETRIEVAL

The storing of samples before and after activation can be accomplished by use of a "library" such as is shown in Fig. 9. It has been found convenient to store 100 samples in a ten by ten matrix equipped with a motorized locator which positions a pneumatic tube outlet over the sample storage library in such fashion that samples can be withdrawn and transported through the automated system on command. The design of such a library requires that care be taken to avoid the accidental crushing and/or breaking of a sample with the resultant contamination of the system. In fact, it follows as a general rule in this sort of development that the entire sample-handling machinery should both avoid contamination and provide a manner in which contamination can be cleaned-up if it occurs accidentally. The use of polyethylene tubing, plastic transfer boxes, and plastic components are all steps in this direction. Furthermore, it has been found useful to employ domestic vacuum cleaners as vacuum sources so that the vacuum-cleaner bag serves as a primary filter of particular radio-activity material if sample breakage occurs.

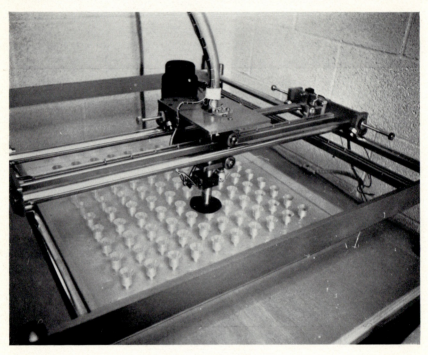

Fig. 9. Storage "library" for samples.

7. NOISE CONTROL

A complex automated activation analysis system, such as the Mark II system, contains several thousand transistors, tubes, and miles of wiring. It is difficult, in a system of this type, to avoid inducing extraneous electronic "noise" into the counting section. The noise can arise as induced electrical signals which are amplified and appear as pulses, or other spurious electrical signals which distort the size, shape, or number of pulses received in the analyzer memory. The avoidance of noise can be accomplished by physically isolating the detector-preamplifier-amplifier stages from the rest of the system in order to insure that the input to the linear amplifiers is as free as possible from noise. The use of co-axial cables is also recommended and the use of modern solid state detectors in place of sodium iodide detectors (which will be discussed further later in this chapter) is particularly notable in this context because the low noise field effect transistor preamplifiers and carefully designed linear amplifiers reduce noise levels in the detector-preamplifier-amplifier complex. Even so, co-axial cables in the solid state spectrometer are always kept very short to reduce noise to a minimum.

8. DRIFT CONTROL

One of the major problems associated with γ-ray spectrometry arises from the "drift" of the spectrometer during the counting interval. Figure 10 illustrates the two most common types of drift. The first type, (curves A and C), occurs in those systems in which the amplifier gain stays constant, though the spectrometer baseline drifts. The second type of drift (curves A and B) involves a situation where the calibration (keV/channel) changes. Fite *et al.* (1961a) have developed a method for drift control which makes use of a small source of plutonium imbedded in a sodium iodide crystal producing a pseudo γ-ray due to the alpha particles from plutonium which appears as a mono-energetic peak at approximately 3 MeV. The pseudo γ-ray peak is monitored electronically, and the spectrometer is continuously automatically adjusted by a drift controller in gain and in baseline position to maintain the pseudo γ-ray peak in its preselected channel. In order for the drift control device to be effective even in the low energy region, it is also useful to have a reference peak, such as the ^{203}Hg low energy γ-ray (≈ 0.3 MeV), which will occur in a low numbered channel. The two reference energy channels—^{239}Pu in the high energy range, and ^{203}Hg in the low energy range—can both be monitored simultaneously. It has been shown that drift control of this type can work quite effectively and for long periods of time. An important problem involving drift control based on a high energy pseudo γ-ray, which has been noted, is that the drift due to temperature variations in the photomultiplier crystal complex and drift in the pseudo γ-ray are inverse; that is, for a given change in temperature, the system drifts in one direction while the pseudo γ-ray effective energy position drifts in the opposite direction. This requires that the entire γ-ray spectrometer be maintained at constant temperature. A problem related to the use of a low energy γ-ray, such as ^{203}Hg, is that other low energy γ-rays can be obscured by the ^{203}Hg peak.

Another source of drift arises from changes in the high voltage supply to the photomultiplier tube. The variation can be avoided by the substitution of high voltage batteries for the electronic high voltage power supply. Batteries are practical for this application because the current drain required to power a photomultiplier tube is so very small that it is comparable to shelf drainage of the battery.

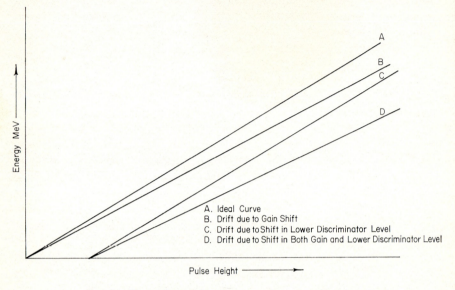

FIG. 10. Illustration of the two types of spectrometer drift.

B. Data Production Considerations

1. READ-OUT FORMAT DESIGN

The data produced in a multi-channel analyzer consist of channel numbers, and the number of counts accumulated per channel. Other information required for quantitative data reduction includes the time at the start of irradiation, the time at the end of irradiation, the time of start of counting, and the "dead-time" during counting, including any changes in "dead-time" during counting. "Dead-time" is that fraction of time during which pulses cannot be accepted because the spectrometer is handling another pulse. In most cases, commercially available multi-channel analyzers do not provide all of this data in the read-out format required by computers, so some of it must manually be obtained. One example of a completely automated read-out is provided in the Mark I Automatic Activation Analysis System shown in Fig. 1. This system provides all of the above information on punched paper tape, coded in such a way as to make it in a format compatable with most modern digital computers. The type of read-out format that is desired by individual experimenters depends to some extent on the quantity of data that they have to use and the method of mathematical data reduction which is being employed. Manual data processing does not necessitate an automated read-out. Computer data reduction and processing requires that clock time information and dead time information accompany the spectral data into the computer in order to provide the computer with the basic information necessary for quantitative calculations.

2. THE TYPES OF READ-OUTS AVAILABLE

The various types of read-outs commercially available include: the character printer, punched paper tape, automatic typewriter, magnetic tape, x–y plotter, and the cathode ray oscilloscope with or without a camera for making a permanent record of the optical output. Even in a highly automated system it is very often desirable to have an x–y plotter and/or a typewriter readout available at all times in order to verify data occasionally during both the calibration and operation of the spectrometer. Usually, however, when one is interested in processing large volumes of data in a routine fashion, the magnetic tape output, in computer-compatible format, is by far the most rapid means of data read-out. By way of comparison, the process of reading out 3,200 channels from a typical multi-channel analyzer on to punched paper tape takes about 40 mins. In the magnetic tape unit, such as is shown in Fig. 11, the entire read-out, including all channel numbers and contents, is completed and transferred to magnetic tape, in computer compatible format, in about one-sixteenth of a second. Such speed is not only useful for the processing of large numbers of samples, but can be most useful where it is necessary to measure the decay characteristics of a rapidly decaying sample by making a number of measurements of the sample during the course of its decay. In many cases, it is useful to count for two or three seconds, read-out the contents of the memory destructively, recount for two or three more seconds, read-out, count for another two or three seconds, and so on. This procedure enables one to obtain a series of spectra which describe the time performance of the various components as a function of decay times.

3. DATA HANDLING PRIOR TO COMPUTER MANIPULATION

A 3,200 channel spectrum consisting of address digits and data digits presents thousands of opportunities for an erroneous digit read-out It is fortunate that the data in adjacent channels provides an indicator of error by calling attention to variations greater than those which can be explained by the counting characteristics of the system. Paper punch machines, punch card machines, and magnetic tape machines all occasionally commit errors or omissions in data read-out. Such a

Fig. 11. Magnetic tape read-out system with 3200 channel analyzer.

development can often cause considerable problems in the reduction of data if unnoticed, but it is possible to program the computer to notice errors or omissions of digits because the spectral discontinuity will be detected by such programs as have been developed by Yule (1969). For example, if 500 counts appear in one channel and 90,000 counts appear in the next channel and 510 counts appear in the next channel, the computer deduces that there is an error in the channel with the 90,000 counts since it is not possible to expect so much change in a single channel without a correspondingly high number of counts in the third channel.

C. Electronic Data Handling and Processing

1. SPECTRUM RESOLUTION CONSIDERATIONS

Typical sodium iodide scintillation detectors have resolutions varying from 7 to 10%. The better the detector resolution, the easier it is to study spectral characteristics. New lithium drifted germanium detectors [Ge(Li)] have resolutions up to 0·13%. However, the total counting efficiency of Ge(Li) is less than that experienced with sodium iodide crystals unless an extremely large (30–50 c.c.) Ge(Li) detector is available. There is substantial hope that in the near future, large solid state detectors, with efficiencies equal to those of sodium iodide detectors, will be available. It is further expected that such detectors will preserve the fine resolutions (4–7 kV wide) of the smaller solid state detector.

2. THE LAW OF LINEAR SUPERPOSITION

This basic law, simply defined says that γ-ray spectra, collected under *proper* conditions may be considered a linear superposition, additively of all of the constituent isotope spectra and background present. From another point of view, in any energy interval, ΔE, one can say that the total number of counts is due to the linear summation of the contributions of all isotopes present. Three commonly used methods of spectral unsorting are spectrum stripping, matrix methods, and least squares curve fitting—all mathematically utilizing the law of linear superposition to resolve a composite spectrum into its parts, and in so doing, all of the methods provide quantitative as well as qualitative information.

3. SPECTRUM STRIPPING METHODS

One of the earliest approaches to spectral resolution was an attempt to "unpeel" or "strip" major spectrum components off, hopefully revealing hidden peaks and other analytically important spectral information in the residual left after the removal of a major constituent. As in the case of all three commonly used computer procedures, the shape of the constituent spectrum due to a single isotope is either measured experimentally and corrected for irradiation conditions and mass, or the spectral shape can itself be prepared by the computer using such methods as those of Heath *et al.* (1967).

The "library" spectrum of a given major constituent identified by its peaks, is first multiplied by some fraction greater than zero and generally less than ten, in order to attempt to "strip" precisely the "correct" contribution of the constituent. If too much is stripped out, the spectrum can reveal this fact by containing counts less than zero in some channels. Successive strippings provide quantization by determining the magnitude of the multiplicative fraction, and comparing it to the mass of the library sample.

The principal short-coming of spectrum stripping is that the statistical and measurement errors add at each stage of stripping, and by the time three or more spectra have been stripped off, the remaining data usually has a very low signal to noise ratio. This difficulty notwithstanding, spectrum stripping is still very interesting because it is the only form of computer data reduction which can be done without any computer additional to the multi-channel analyzer itself which is actually a small, special purpose digital computer. For example, many multichannel analyser manufacturers offer a small magnetic tape unit which can store library data and a multiplier which can vary the magnitude of the library data for the purpose of instrumental spectrum stripping. In this instance, the "subtract" circuits of the analyzer are used to strip out a given component until the peak or peaks due to that component disappear.

4. MATRIX METHODS

Breen *et al.* (1961) investigated the use of the following set of equations to resolve a three component γ-ray spectrum:

$$H_{\Delta E_1} = A_1 e^{-at_1} + B e^{-bt_1} + C_1 e^{-ct_1}$$
$$H_{\Delta E_2} = A_2 e^{-at_2} + B_2 e^{-bt_2} + C_2 e^{-ct_2}$$
$$H_{\Delta E_3} = A_3 e^{-at_3} + B_3 e^{-bt_3} + C_3 e^{-ct_3}$$

Where $H_{\Delta E_N}$ is the total number of counts in the channel corresponding to ΔE_N, and is a measured quantity. A, B and C are functions of the masses of the three constituent isotopes and are not variables. The terms a, b and c represent the decay factors of the three isotopes and are not variables. The term t_n is the mean time during which the spectrum is measured. Thus, if one assumes the qualitative identity of the three components based on the location of spectral peaks, thus fixing the decay factors, the quantization can be performed by solving these three simultaneous equations for the three linear unknowns: A, B and C.

The matrix method has been demonstrated for a small number of components (about 3), but problems related to matrix arithmetic, such as the errors introduced when there is a large difference in the magnitude of matrix terms, makes this method of comparatively little use.

5. LEAST SQUARES CURVE FITTING

Spectral unsorting can be performed, based on Gauss methods, that a complex composite curve could be approximated by its constituent curves added together in such fashion as to minimize the square roots of the variations between the observed spectrum and that spectrum which would result from a combination of library curves all multiplied by a given factor dependent upon mass and irradiation conditions. Through an approximation process, it has been demonstrated that very good results can be obtained by computer data reduction employing the least squares method. Certain restraints, such as the prohibition of negative masses, have been employed by Smith (1966).

D. Modern Trends

The rapid adoption of automation and computer data reduction suggest that the inherent advantages of these procedures provide realistic benefits even at the present state of development. Certain areas of necessary development may, however, be predicted. For example, it will be necessary to have substantially faster analog-to-digital converters if large Ge(Li) detectors (30–50 c.c. or even larger) are combined with analyzers having over 10,000 channels.

It also seems logical to expect that analog computers, especially of the rep-op type, will be vital for on-line solution of the basic time dependent equations of activation analysis. Such a system was described by Fite *et al.* (1961b), but only now are manufacturers producing these devices. The analytical application which is being handled this way is the analysis of oxygen in steel, a situation in which a simple, analog based system has clear advantages over its nearest competitor, vacuum fusion (Van Wyk *et al.*, 1966).

Another development which seems to be very important is the addition of "positional" analysis to qualitative and quantitative analyses. It has been shown (Menon *et al.*, 1966) that a piece of metal could be "mapped" with the location of its impurities being plotted in two-dimensions. The large scale activation analysis system could be used, therefore, to either measure many points on a single piece of material, in order to establish compositional profiles.

References

Breen, W. M., Fite, L. E., Gibbons, D., Wainerdi, R. E. (1961). *Trans. Am. nucl. Soc.* **2**, 2, 244

Covell, D. F. (1959). *Analyt. Chem.* **31**, 1785.

De Hevesy, G. and Levi, H. (1936). *K. danske Vidensk. Selsk. Skr.* **14**, No. 5.

Fite, L. E., Gibbons, D. and Wainerdi, R. E. (1961a). *In* "Modern Trends in Activation Analysis," pp. 102–107. College Station, Texas.

Fite, L. E., Gibbons, D. and Wainerdi, R. E. (1961b). Computer Coupled Automatic Activation Analysis, Report TEES 2671-1. Department of Commerce.

Gibbons, D., Fite, L. E. and Wainerdi, R. E. (1962). "Analytical Chemistry," pp. 269–273. Elsevier, Amsterdam.

Heath, R. L., Helmer, R. G., Schmittroth, L. A., Cazier, G. A. (1967). *Nucl. Instrum. Meth.* **47**, 281.

Kuykendall, W. E. Jr. and Wainerdi, R. E., (1960). *In* "Use of Radioisotopes in the Physical Sciences and Industry." IAEA Vienna, Vol. II, pp. 233–41.

Menon, M. P. and Wainerdi, R. E. (1965). *In* "Modern Trends in Activation Analysis," pp. 152–156. College Station, Texas.

Menon, M. P., Rainosek, A. P. and Wainerdi, R. E. (1966). *Nucl. Appl.* **2**, 335–340.

Nuclear Science Series Monographs, Radiochemistry of the Elements (NAS–NS 3001–3058) and Radiochemical Techniques (NAS–NS 3101–3110).

Salmon, L. (1961). UKAEA Report, AERE, R–3640.

Schumann, R. and McMahon, J. (1956). *Rev. scient. Instrum.* **27**, 675.

Smith, L. H. (1966). *Analytica chim. Acta* **36**, 149–165.

Van Wyk, J. M., Cuypers, M. Y., Fite, L. E. and Wainerdi, R. E. (1966). *Analyst, Lond.* **91**, 316–323.

Yule, H. P. (1969). "A Computer Program for Analysis of Activation Analysis Gamma-ray Spectra" to be published in the Journal of Radioanalytical Chemistry.

STANDARD MATERIALS AND INTERCOMPARISONS

H. J. M. BOWEN

Chemistry Department, The University, Reading, Berks.

Although activation analysis is theoretically an absolute method of analysis, in practice it always involves comparison with a known standard of a pure element. The preparation, canning and handling of such standards merit a chapter of their own for discussion. Another kind of standard that can be used is a homogeneous material of known composition. This kind of standard forms the subject matter of the present chapter.

The importance of standard materials is that they can be used to measure the *accuracy* of a method of analysis. The accuracy of any experimental result refers to its closeness to the truth. It should not be confused with the *precision* of any set of results, which is simply a measure of their internal consistency. Accuracy and precision are quite independent quantities. For example, the Babylonians who apparently always found the value of π to be 3, were quite precise but rather inaccurate. The modern literature of analytical chemistry is filled with examples of work which is either precise but inaccurate, or accurate but imprecise. Truly precise, accurate work is as rarely seen as are imprecise, inaccurate results; the latter are mostly thrown out by the referees.

A good way of testing both the accuracy and precision of a particular technique or set of techniques is the intercomparison method, sometimes known as a "Round Robin". This involves sending one or more standard materials for analysis to workers in many different laboratories, and comparing the results produced. Standard statistical procedures can then be used to determine the "best" mean, and hence the accuracy of individual workers or techniques (cf, e.g. Fairbairn and Schairer, 1952; Kenworthy *et al.*, 1956; Ward and Heeney, 1960; Cook *et al.*, 1963).

There is nothing particularly novel about standards, intercomparisons or the statistics involved. If we consider only standards for elementary analyses, the first materials appear to have been prepared in 1909–10 by the U.S. Bureau of Standards, and by Ridsdale and Ridsdale in Britain in 1916. The first standard materials were steels, iron and manganese ores, and for many years these samples have been of great use to analysts in the metallurgical industry. At the present day it is possible to buy samples of ferrous alloys in which a number of elements have been determined. Most of these elements are present as major constituents, and although ferrous alloy standards are available for 28 elements, relatively few standards have been analyzed for more than ten elements. Similar standards are now commercially available for alloys of aluminium, copper, lead, magnesium, nickel, tin and zinc, as well as for ores, slags and refractory materials. Despite the fact that at least 70 elements could probably be determined in these standards by activation analysis, very little work has been reported on their trace constituents.

The first serious attempt to produce standard materials in which both major elements and trace constituents could be determined was in the geochemical field (Fairbairn, 1951). Fairbairn and his collaborators ground up 50–100 kg of two rock samples, a granite G–1 and a diabase W–1, to pass an 80-mesh sieve. Some contamination was probably introduced during the grinding process, but this should not have affected subsequent intercomparisons provided that it was homogeneously distributed. One unanticipated problem in handling rock powders is that of separation of particles of different density. If the powder is stored in bulk, minute vibrations cause settling of the denser particles to the bottom of the storage container, so that the whole must be thoroughly mixed before it is divided into smaller samples (Roubault, 1963). It now appears that not all the samples of G–1 that were distributed were identical (Flanagan, 1960).

Despite these difficulties, the appearance of G–1 and W–1 started a revolution in the accuracy of geochemical analysis. Initially 100 g samples were distributed to 34 analysts in 25 different laboratories, and when the results of the first intercomparison were published, demand for the standards grew to such an extent that within 15 years the original supplies of G–1 were practically exhausted. Thus while in 1951 analyses were only provided for 25 elements by a maximum of 11 analysts per element, by 1965 analyses were available for 77 elements provided by over a hundred laboratories (Fleischer, 1965). The analytical methods used have included activation analysis, atomic absorption, colorimetry, flame photometry, gravimetry, isotope dilution, mass spectrometry, polarography, spectroscopy, volumetry and X-ray fluorescence. Some 59 of the elements in these standard rocks have now been determined by activation analysis.

The results accumulated in this way have had a profound effect on the development of geochemistry. They have helped to increase both the precision and the accuracy of geochemical analysis, and have led to important and exciting developments in cosmochemistry. With regard to precision, it is instructive to compare early and recent data for copper and strontium in G–1 and W–1. Each datum given in Table I represents

TABLE I

Copper and Strontium in G–1 and W–1*

Element	Rock	Date of Analyses	Data (p.p.m.)	Mean ± s.d.
Cu	G–1	1951	8, 8, 15, 20, 50	20·2 ± 17·4
Cu	G–1	1962–5	7·8, 10, 10, 11, 11, 11, 11·5, 12, 12, 13, 14, 20	11·9 ± 3·0
Cu	W–1	1951	44, 80, 88, 128, 128	94 ± 36
Cu	W–1	1962–5	80, 103, 109, 110, 110, 110, 110, 110, 110, 115, 118, 118, 120, 124, 130, 130	113 ± 12
Sr	G–1	1951	118, 254, 930	434 ± 435
Sr	G–1	1962–5	191, 226, 250, 250, 253, 260, 262, 265, 265, 271, 307, 308, 310	263 ± 33
Sr	W–1	1951	144, 254, 422	273 ± 140
Sr	W–1	1962–5	151, 155, 156, 160, 169, 175, 176, 177, 178, 180, 199, 200, 210, 210, 227, 227	184 ± 25

*Data from Fairbairn (1951) and Fleischer (1965), neglecting aberrant mass spectrometric results of Brown and Wolstenholme (1964).

the mean value obtained by a particular laboratory. It can be seen that the overall precision, or agreement between different analysts, has greatly improved in the last decade.

With regard to accuracy, the most recent data of Fleischer indicate that agreement between results by different techniques is reasonably good in nearly all cases, despite a spread of roughly ±20% between results by individual analysts. The following exceptions should be noted:

1. For the lanthanides, Ce, La, Nd and Yb, recent activation and mass spectrometric analyses give results lower by a factor of 2–3 than those found by spectroscopy.
2. For cobalt, activation and X-ray fluorescence give results for W–1 that are 25% higher than those found by spectroscopy and by colorimetry.
3. For fluorine and molybdenum, single determinations by activation have given results 50–100% higher than those found by other techniques.
4. For both iridium and osmium, two independent workers have obtained results differing by an order of magnitude using activation.

One result of the success of G–1 and W–1 as rock standards has been the development of their own successors. These have been discussed by Taylor and Kolbe (1964), who conclude that many more properly

TABLE II

Desirable Rock Standards

Type	Available at present	Source
Iron meteorite	—	—
Chondrite	—	—
Ultrabasic rock	—	—
Basalt/Diabase	W–1	U.S. Geol. Survey
Intermediate rock	Sy–1 Syenite	McGill Univ., Canada
Granite	G–1	U.S. Geol. Survey
	USBS 4981 & 4983	U.S. Nat. Bur. Stds.
Limestone	USBS 1A	U.S. Nat. Bur. Stds.
Shale	USBS 98 plastic clay	U.S. Nat. Bur. Stds.
	BCS 269 firebrick	Bur. Analysed Samples, U.K.
Sandstone	USBS 102 silica brick	U.S. Nat. Bur. Stds.
	BCS 267 silica brick	Bur. Analysed Samples, U.K.
Tektite	—	—

prepared rock and mineral standards are needed. The 10 rock standards suggested by Taylor and Kolbe are listed in Table II. Flanagan (1967) gives results for 35 elements in six new silicate rock standards, and further sources of standard rocks are listed by Flanagan and Gwyn (1967).

Taylor and Kolbe also give analytical results for 26 elements in these and other possible standards.

Cook *et al.* (1963) have published the results of an intercomparison by nine laboratories of a standard aqueous solution containing about

5 p.p.m. Cu, 16 p.p.m. Cr, 15 p.p.m. Hg and 4 p.p.m. Mn. Their results for four techniques can be summarized by comparing the standard deviations of a single determination at the 5% level, as listed in Table III.

TABLE III

Standard Deviation of a Single Determination as Percentage of the Mean

Technique	Cu	Cr	Hg	Mn
Activation	28·1	11·2	17·8	17·0
Colorimetry	6·2	7·9	10·5	10·0
Polarography	15·5	12·7	13·5	37·2
Spectroscopy	42·9	19·1	46·1	34·4

TABLE IV

Some Biological Standard Materials

Sample	Name of species		Source
Dry leaf	*Citrus aurantia*	(Orange)	A. L. Kenworthy, East Lansing, Mich., U.S.A.
Dry leaf	*Malus sylvestris*	(Apple)	
Dry leaf	*Prunus persica*	(Peach)	
Dry leaf	*Pyrus communis*	(Pear)	
Dry leaf	*Brassica oleracea*	(Kale)	H. J. M. Bowen, Reading Univ., U.K.
Flour	*Triticum vulgare*	(Wheat)	I.A.E.A., Vienna, Austria
Flour	*Oryza sativa*	(Rice)	
Bone	*Bos taurus*	(Cow)	
Dried Milk	*Bos taurus*	(Cow)	
Dried Blood	*Homo sapiens*	(Man)	
Dried Serum	*Homo sapiens*	(Man)	
Dried Serum	*Homo sapiens*	(Man)	Dade Reagents, Miami, Florida, U.S.A.
Mollusc shell	*Anodonta corpulenta*		M. Merlini, Ispra, Italy
Mollusc shell	*Elliptio crassidens*		
Mollusc shell	*Mercenaria mercenaria*		
Mollusc shell	*Quadrula pustulosa*		

In 1964 I pointed out that there were few or no standard biological materials available, and suggested some possible standards (Bowen, 1964). Since that time several standard biological materials have become available, and a selection of these are given in Table IV. Biochemical standards and their limitations have been discussed by Radin (1967).

The preparation, properties and homogeneity of the kale powder have now been described (Bowen, 1965 and 1966). Since very few results have as yet been reported for this or any other biological material, the rest of this chapter will discuss the results to date. Thirty-four laboratories have submitted analyses, mostly in quadruplicate, and 44 elements have now been determined in this material. A statistical review of the data reveals numerous inconsistencies, but shows that certain techniques give consistently high or low results for certain elements (Bowen, 1967).

The kale powder is slightly hygroscopic and contains about 5% of water when despatched. The drying procedures used by different analysts differ slightly, and may have caused errors of $\pm 1\%$ in results calculated in terms of dry weight. It is recommended that samples of the material should be dried for 20 hr at 90°C, and subsequently cooled in a desiccator, for determination of dry weight.

Table V summarizes the data received so far for the elementary analysis of kale powder. The data are given in μg/g dry weight, as means of results from each laboratory, keeping techniques of analysis separate. They are followed by the number of replicate analyses in brackets. Techniques are abbreviated as follows: act = activation analysis; ata = atomic absorption spectroscopy; bet = beta counting; cat = catalytic technique; col = colorimetry; fla = flame photometry; flu = fluorescence analysis; gra = gravimetry; γ = gamma spectroscopy; iso = isotope dilution; pol = polarography; spe = spectrometry: tur = turbidimetry; vol = volumetric analysis; X-ray = X-ray spectrometry.

Since most of the data in Table V have been analyzed statistically elsewhere (Bowen, 1967), the discussion will be confined to the results obtained by activation analysis. The inter-laboratory precision of this technique is good for As, Au, Br, Ca, Co, Cu, Hg, K, Mn, P, Rb, Sb, Sc, Sr, W and Zn, but poor for Ag, Al, Cd, Ce, Cl, Cr, Ga, Ir, Mo and Se. This situation will doubtless change when more results are available.

The accuracy of the technique is in doubt in the case of Al, As, Hg, I and K. The aluminium analyses are difficult in view of the short half-life of ^{28}Al and the many interfering activities. The same is true of iodine, where the discrepancy between results by activation and catalytic methods may be due to ^{38}Cl interfering with the determination of ^{128}I. The self-consistency of the activation results for arsenic and mercury suggest that the single colorimetric analysis is erroneous in each case. In the case of potassium, there is a marked discrepancy between the results obtained by four laboratories using γ-spectrometry and those obtained by seven other techniques which are mutually consistent. It

TABLE V

Summary of Analyses of Standard Kale

Element	p.p.m. (No. of replicates)	Technique	Remarks	Best value p.p.m.
Ag	0·03(3), 0·46(1) <0·45(1), 0·5(1)	act act(γ)	agreement poor	0·03?
Al	7·4(4), 78·5(2), 38·7(4), 41(1) 35·1(4), 35·1(4), 27·3(4), 44·7(4), 6·4(3) 69·7(4), 88·4(5)	act(γ) col spe	methods disagree	37
As	0·16(4), 0·11(8), 0·12(4), 0·22(2), 0·12(6), 0·14(1), 0·18(2) 1·8(4)	act col	methods disagree	0·14
Au	0·0021(3), 0·0022(4), 0·0029(2)	act		0·0023
B	21·1(8), 46·8(6), 49·5(4), 49·6(5), 56·0(4), 55·0(4), 52·3(4), 44·0(4), 56·3(4), 38·9(3), 39·6(3) 41·2(5)	col spe	precision poor	44
Ba	4·1(4) 5·1(4), 4·0(4)	act spe		4·4
Br	25·4(4), 24·9(3), 24·5(5), 23·0(6), 29·5(15)	act		27
Ca	40400(1), 39000(3), 40000(2), 34100(4), 38600(4), 42000(4) 43200(4), 41300(4), 43900(4), 42800(4), 41100(4), 38300(6), 42500(1), 25800(1) 41000(2) 31200(4), 45800(4), 39900(5), 42900(4), 30750(4) 39400(4), 40300(2), 44000(3), 43200(4) 42400(2), 40000(1)	act ata col fla vol X-ray	last ata result is erroneous	40000
Cd	0·38(4), 0·63(5), 0·91(6) 1·0(4)	act pol	precision poor	0·74
Ce	0·18(4), 0·46(4)	act		0·3?
Cl	4450(4), 2475(4), 3270(7), 2180(1), 2300(1), 3880(4), 3980(1) 3920(5) 2400(1)	act vol X-ray	precision poor	3450

TABLE V (cont'd)

Element	p.p.m. (No. of replicates)	Technique	Remarks	Best value p.p.m.
Co	0·052(3), 0·052(5), 0·060(4), 0·081(2)	act	second spe value erroneous	0·056
	0·05(1), 0·065(4), 0·055(4), 0·058(4), 0·06(1)	col		
	0·075(4), 2·0(5)	spe		
Cr	0·42(4), 0·18(4)	act	precision poor	0·33
	0·73(1)	ata		
	0·29(4)	spe		
Cs	0·069(6), <0·06(1)	act		0·069?
Cu	4·4(4), 3·6(2), 4·1(6), 4·6(15), 5·6(1), 6·5(6)	act		5·0
	5·8(4), 5·4(5), 5·1(4), 6·0(1), 4·7(4), 6·3(4), 5·2(2), 6·3(7)	ata	ata gives high results	
	3·7(4), 4·7(5), 4·9(4), 4·3(4), 4·8(4), 4·8(4), 4·3(4), 4·6(3), 5·4(4), 4·3(4), 5·2(4), 4·4(4), 4·3(4), 6·5(4), 5·3(2), 9·5(5), 17(3)	col	last two col values erroneous	
	4·5(1), 5·5(4)	pol		
	6·4(4), 4·6(5)	spe		
Dy	<0·024(1)	act		
F	5·55(4), 4·23(3)	vol		5·0
Fe	155(1), 120(7), 101(2), 113(1)	act		120
	112(4), 134(5), 106(4), 117(4), 143(7), 149(3), 59(1)	ata		
	157(4), 125(4), 111(6), 120(6), 114(4), 99(4), 105(4), 115(4), 155(5), 106(4), 110(4)	col		
	145(4), 84(5)	spe		
	90(1)	X-ray		
Ga	0·064(3), 0·027(3)	act	agreement poor	0·05?
Hf	<0·07(1)	act		
Hg	0·14(5), 0·16(4), 0·18(4), 0·15(4), 0·165(2), 0·17(9), 0·18(3)	act	methods disagree	0·16
	0·012(4)	col		
I	0·27(2), 0·26(2)	act	methods disagree	0·080?
	0·11(4), 0·068(4), 0·063(4)	cat		

Table V (cont'd)

Element	p.p.m. (No. of replicates)	Technique	Remarks	Best value p.p.m.
In	<0·3(1)	act		
Ir	<0·021(1), <0·0005(1), <0·013(4)	act	agreement poor	
K	26000(3), 22000(4), 23800(4), 24050(1)	act		25000
	15540(5), 21900(5), 20600(1), 16300(1)	act(γ)	act(γ) gives low results	
	24800(4), 21700(1), 23700(6)	ata		
	26600(4), 33200(1)	bet		
	26900(11), 23000(4), 24400(4), 24300(2), 23400(4), 23900(5), 23600(4), 24900(9), 22500(1), 25350(4), 22900(4), 24700(8), 24700(4), 24000(1)	fla		
	29300(4)	iso		
	25600(5)	spe		
	23000(1), 24800(1), 24800(2)	X-ray		
La	0·077(6)	act		0·077?
Mg	1350(6), 1605(1)	act(γ)		1600
	1670(4), 1600(4), 1610(4), 1700(5), 1570(4), 1600(1), 1560(9), 1640(6), 1650(1)	ata		
	1540(4), 1530(4)	col		
	3460(5)	fla	too high	
	1140(5)	spe	too low	
	1700(4), 1450(3), 1550(4)	vol		
Mn	12·6(3), 13·7(8), 14·6(2), 16·9(4), 14·6(6), 16·0(1), 14·9(10), 14·0(4), 14·4(1), 17·5(1)	act		14·9
	18·0(4), 14·4(6), 13·9(4), 15·0(1), 14·7(4), 16·7(4), 15·6(4), 17·9(7)	ata		
	13·9(6), 12·2(4), 14·6(2), 17·0(4), 11·5(4), 15·8(4), 15·0(4), 16·5(4), 15·5(4), 15·0(3), 10·0(2), 29·0(4), 25·0(5)	col	last two col. values erroneous	
	10·0(5)	spe	too low	
	13·0(1)	X-ray		
Mo	3·1(4), 1·7(5), 2·6(10), 1·45(2)	act		2·3
	2·8(7), 2·4(4), 1·9(2), 2·25(4), 2·2(4), 1·95(4), 2·05(2), 2·3(8), 2·0(4)	col		
	2·2(1)	pol		
	1·5(4), 0·59(5)	spe	both too low	

TABLE V (cont'd)

Element	p.p.m. (No. of replicates)	Technique	Remarks	Best value p.p.m.
N	43500(2)	act		43200
	42500(5)	spe		
	42600(4), 42400(4), 43300(4), 41100(2), 43400(4), 44100(10), 45800(4), 41300(4)	vol		
Na	1930(10), 2160(5)	act		2500
	2500(5), 2640(1)	act(γ)		
	2280(4), 2170(4), 2910(1), 2525(1), 2970(6)	ata		
	2930(14), 2980(4), 3250(5), 2430(9), 2600(4), 1920(4)	fla	fla results significantly high	
	1220(5)	spe	too low	
Ni	11·0(4)	ata	methods disagree	
	2·6(4)	spe		
O	51500(1)	act		51500
P	4410(4), 4530(4), 4590(1)	act		4500
	4460(6)	ata		
	4560(6), 4320(4), 4430(4), 4300(2), 4600(4), 4660(5), 4810(4), 4500(9), 4740(4), 4490(4)	col		
	4020(5)	spe	low	
	4400(1)	X-ray		
Pb	1·6(4), 3·8(5)	col	precision poor	3·2
	5·4(4), 3·0(4)	pol		
	2·1(4)	spe		
Rb	56(4), 50(5), 57·5(1)	act		53
	49(1), 57(2)	X-ray		
Re	<0·3(1)	act		
Ru	0·0045(2)	act		0·0045?
S	17300(5), 14300(8)	col		16000
	19800(5)	gra		
	24900(4)	tur	erroneous?	
	13500(4)	vol		
	13700(1), 18000(2)	X-ray		

TABLE V (cont'd)

Element	p.p.m. (No. of replicates)	Technique	Remarks	Best value p.p.m.
Sb	0·065(6), 0·099(4), 0·059(8)	act		0·070
Sc	0·0088(2), 0·0079(2), 0·0087(6)	act		0·0086
Se	0·017(4), 0·15(8), 0·046(2), 0·62(6) 0·14(4), 0·14(4)	act flu	agreement poor	0·15?
Si	241(2) 243(3) 130(1)	act col X-ray		240
Sm	0·16(4)	act		0·16?
Sn	0·16(4)	spe		0·16?
Sr	75(4), 76(2) 78(6), 101(8) 150(5) 65(1), 101(2)	act ata spe X-ray	too high	98
Ta	<0·1(1)	act		
Th	0·0092(7)	act		0·0092?
Ti	0·33(4) 2·75(4)	col spe	methods disagree	
Tl	0·15(4)	col		0·15?
U	0·0135(4), 0·0081(4)	act		0·01
W	0·064(4), 0·057(4)	act		0·06
Zn	33·5(4), 30·3(2), 32·1(7), 32·8(4), 37·5(1), 26·7(2), 34·6(8) 35·3(6), 33·8(4), 32·6(4), 35·6(5), 32·8(4), 33·0(1), 34·3(4), 38·0(4), 32·0(1), 32·2(10), 35·6(7) 27·8(4), 38·0(3), 23·7(3), 35·5(4), 31·5(4), 34·0(3), 24·0(3) 36·0(2), 20·0(5), 31·0(1) 33·6(5) 29(1), 35(2)	act ata col pol spe X-ray		32
Zr	<20(1)	act		

appears that there is some unexplained systematic error in γ-spectrometry of potassium, which may be connected with the poor resolution of the 1·52 MeV peak of ^{42}K from the 1·37 MeV peak of ^{24}Na in irradiated biological material, and which gives rise to low results.

The sensitivity of the technique is illustrated by the fact that it is the only method so far applied to determine Ag, Au, Br, Ce, Cs, Ga, La, Ru, Sb, Sc, Sm, Th, U and W in this material. The upper limits for Dy, Hf, In, Ir, Re, Ta and Zr have been determined by γ-spectrometry without chemical separation.

Acknowledgments

I wish to thank the many analysts who co-operated in analyzing samples of standard kale material and reporting their results, namely: D. A. Beardsley, G. B. Briscoe, J. Ruzicka and M. Williams of Birmingham, England; E. G. Bradfield of Bristol, England; R. B. Carson of Ottawa, Ontario; P. A. Cawse of Wantage, England; R. F. Coleman of Aldermaston, England; R. E. Collier of Derby, England; G. B. Cook of Vienna, Austria; G. C. Cotzias and S. T. Miller of Brookhaven, New York; F. Cox of Raleigh, N. Carolina; J. Cuypers and R. E. Wainerdi of College Station, Texas; D. J. David of Canberra, Australia; E. B. Davies and K. J. McNaught of Hamilton, New Zealand; D. J. Fricker of Ashford, Kent, England; F. Girardi of Ispra, Italy; K. E. Heinerth of Dusseldorf, Germany; R. A. Howie, H. D. Livingston and H. Smith of Glasgow, Scotland; E. Jackson of Leeds, England; E. J. Jewell of Exeter, England; J. B. Jones of Wooster, Ohio; J. C. Lane of Wexford, Ireland; J. K. Miettinen and H. Puumala of Helsinki, Finland; D. F. C. Morris and J. C. Gupte of London, England; B. R. Petersen of Copenhagen, Denmark; E. E. Pickett of Columbia, Missouri; E. Poulsen of Lyngby, Denmark; K. Samsahl of Studsvik, Sweden; B. Sjöstrand and T. Westermark of Stockholm, Sweden; T. Teichman of Ottawa, Ontario; J. F. C. Tyler of London, England; G. M. Ward of Harrow, Ontario; A. M. Williams of Cardiff, Wales, Britain; T. Williams of Reading, England; H. P. Yule of San Diego, California; D. A. Becker and G. W. Smith of Washington, D.C., U.S.A.; K. P. Champion and R. N. Whittem of Sutherland, N.S.W., Australia; B. L. Hampson of Lowestoft, England; K. Heydorn of Risö, Denmark; J. Hoste, A. Speecke and P. Van den Winkel of Ghent, Belgium; C. K. Kim of Columbia, Missouri; L. Kosta of Ljubljana, Yugoslavia; E. L. Kothny of Berkeley, California; O. Landström of Stockholm, Sweden; G. E. Likens of Hanover, N.H., U.S.A.; L. O. Plantin of Stockholm, Sweden; P. Schramel of Munich, Germany; and H. Van Brandt of Vietnam.

References

Bowen, H. J. M. (1964). *In* "L'Analyse par Radioactivation et ses Applications aux Sciences Biologiques", (D. Comar, ed.), p. 199. Presses Universitaires de France, Paris.
Bowen, H. J. M. (1965). *In* "Proceedings of the SAC Conference at Nottingham", (P. W. Shallis, ed.), p. 25. Heffer, Cambridge.
Bowen, H. J. M. (1966). *In* "Modern Trends in Activation Analysis", p. 58. College Station, Texas.
Bowen, H. J. M. (1967). *Analyst, Lond.* **92**, 124.
Brown, R., and Wolstenholme, W. A. (1964). *Nature, Lond.* **201**, 598.
Cook, G. B., Crespi, M. B. A. and Minczewski, J. (1963). *Talanta* **10**, 917.
Fairbairn, H. W. (1951). *U.S. Geol. Survey Bull.* **980**, 1.
Fairbairn, H. W., and Schairer, J. F. (1952). *Am. Min.* **37**, 744.
Flanagan, F. J. (1960). *U.S. Geol. Survey Bull.* **1113**, 113.
Flanagan, F. J. (1967). *Geochim. Cosmochim. Acta* **31**, 289.
Flanagan, F. J., and Gwyn, M. E. (1967). *Geochim. Cosmochim. Acta* **31**, 1211.
Fleischer, M. (1965). *Geochim. Cosmochim. Acta* **29**, 1263.
Kenworthy, A. L., Miller, E. J., and Mathis, W. T. (1956). *Proc. Am. Soc. Hort. Sci.* **67**, 16.
Radin, N. (1967). *Clin. Chem.* **13**, 55.
Roubault, M. (1964). *In* "L'Analyse par Radioactivation et ses Applications aux Sciences Biologiques", p. 3.
Taylor, S. R. and Kolbe, P. (1964). *Geochim. Cosmochim. Acta* **28**, 447.
U.S. Bureau of Standards, Circulars **14** (1909), **25** (1910) and **26** (1910).
Ward, G. M. and Heeney, H. B. (1960). *Can. J. Plant Sci.* **40**, 589.

ACTIVATION ANALYSIS TECHNIQUES FOR DOSIMETRY AND HAZARD CONTROL WITH HIGH ENERGY RADIATIONS

L. SKLAVENITIS

Ingenieur au Commissariat a l'Energie Atomique, Centre d'Etudes Nucleaires de Saclay, France

A. Introduction	115
B. Activation by High Energy Radiations	116
1. Definition of high energy radiations	116
2. Cross sections for high energy reaction	117
3. The use of high energy reactions for particle fluence assessment	118
C. In-beam Fluence Measurements	118
1. Choice of the monitor reactions	119
2. Some reactions used to monitor pure particle beams	119
3. Dose calculation	124
D. Composite Radiation Fields	125
1. Measurements in mixed radiation fields	125
2. Dosimetry within a phantom	127
E. Calculation of the Absorbed Dose	130
F. Possibility of Use of Activation Detectors as Particle Dosimeters	131
G. Conclusions	133
References	134

A. Introduction

Activation analysis has widely been used for fluence assessment and for dosimetry of thermal and fast neutrons produced by nuclear reactors; the activity induced in a detector, usually by (n, γ) reaction, is proportional to the neutron fluence.

The same principle can also be applied in studying the hazards associated with exposure to high energy electromagnetic radiation or particles. Such radiations (protons, neutrons and sometimes mesons and photons of sufficiently high energy) are produced in the vicinity of particle accelerators. In what follows a brief review will be given of activation techniques used for assessment and hazard control of high energy radiations.

B. Activation by High Energy Radiations

1. DEFINITION OF HIGH ENERGY RADIATIONS

Primary particles accelerated by high energy machines produce secondary radiations by interaction with the various targets, including parts of the machine and shield. The nature and energy spectrum of these radiations depend on many parameters, including the nature and energy of the primary particles and the chemical composition of the targets.

For ions (usually protons) and electrons with energy in the GeV region, the secondary radiations in the target neighbourhood consist of cascade nucleons of energy several hundred MeV and evaporation nucleons with energy of several MeV.

In general, a high energy particle interacting with a nucleus initiates a cascade of intranuclear collisions (Hudis and Miller, 1959) leading to instantaneous emission of several energetic nucleons. The residual nucleus is left in a highly excited state and may emit a number of lower energy nucleons or fragments; these are called evaporation particles.

Besides the above mentioned nucleons, nuclear reactions initiated by particles whose energy exceeds about 300 MeV, create, by radiative collisions, mesons and strange particles.

Moreover there is production of γ-rays resulting from nuclear reactions, nuclear de-excitation, electron annihilation and π° disintegration. These photons produce in their turn photoneutrons and, by materialization, electron-positron pairs.

In the case of electron accelerators there is mainly production of X-rays by the bremsstrahlung process and subsequently photoneutron emission from target nuclei.

The nature and energy spectra of high energy radiations change with distance from the target. Monte Carlo methods have been used for the study of radiation spectra emitted from target nuclei of various atomic numbers, bombarded by energetic particles (Metropolis *et al.*, 1958; Wallace and Sondhaus, 1962; Dostrovsky *et al.*, 1958; Alsmiller *et al.*, 1967; Bertini, 1963a,b, 1965) as well as those produced and propagated within material targets (Moyer, 1961; O'Barrell, 1964), thus completing previous studies (Hess *et al.*, 1959; Rossi, 1956).

In the region outside the shield, neutrons and γ-rays predominate, accompanied by weak interacting particles such as μ-mesons.

The high energy radiations with which we shall be concerned here include nucleons and mesons of energies extending from a few MeV to several GeV and also photons of energy corresponding to the giant resonance region.

2. CROSS SECTIONS FOR HIGH ENERGY REACTIONS

In the energy region above a few MeV most of the previously quoted radiations interact with nuclei through various nuclear reactions producing activation of the irradiated material.* These reactions can be broadly divided into two classes according to the way in which cross section varies with energy.

(a) Relatively simple reactions, such as (p, n), (n, p), (p, pn), present a threshold at a few MeV; cross sections increase rapidly to attain a maximum, of the order of 100 millibarns at a few tens of MeV, then decrease steadily with increasing energy, to a small, almost constant value.

(b) Spallation reactions yielding product nuclei very different from the target nucleus—for example, ^{197}Au$(p,$ spallation$)^{149}$Tb. In this class, thresholds are generally higher. Cross sections increase less rapidly to a plateau between a few millibarns and a few tens of millibarns and remain more or less constant thereafter.

Cross sections at high energies are little influenced by the nature of the bombarding nucleon. For this reason it is sometimes possible to take as the cross section of a spallation reaction, initiated by neutrons, the corresponding value of the reaction leading to the same product and initiated by protons. Such examples are the reactions ^{27}Al$(p, 5p5n)^{18}$F and ^{27}Al$(n, 4p6n)^{18}$F.

The most accurately known cross sections are those for reactions initiated by protons; some values are known with accuracy of 3–5%, as for example the proton reaction on graphite, ^{12}C$(p, pn)^{11}$C. Other cross sections are established relatively to better known ones and thus the error is usually larger. In some cases discrepancies are observed between two sets of values given by different authors. In spite of these inaccuracies cross sections for many reactions, as a function of proton bombarding energy, are well known (Bruninx, 1961, 1962).

The situation is less satisfactory for neutron-initiated reactions. Reasonably accurate values are available up to 14 MeV, with some information at energies up to about 20 MeV. Values for higher energies

---

* One should notice that this is not the case for particles like electrons and μ-mesons undergoing weak interactions.

are extremely scarce; moreover errors become very important because of neutron energy uncertainties.

As far as mesons are concerned, π-mesons react very much like protons and are usually treated as such. For bombarding energies up to a few GeV, π-meson fluences within a target are much smaller than nucleon fluences. Except for the $^{12}C(\pi-, \pi-n)^{11}C$ reaction (Reeder, 1962; Poskanzer et al., 1961), cross section data for meson reactions are meagre. Reactions of both of the classes mentioned on p. 117 can be used for monoenergetic particle fluence assessment, provided that the appropriate cross section is known.

For polyenergetic particles presenting an energy spectrum, a mean cross section averaged over the spectrum is necessary. The mean value of the cross section is strongly dependent on energy spectrum for class (a) and independent of it in class (b) reactions.

3. THE USE OF HIGH ENERGY REACTIONS FOR PARTICLE FLUENCE ASSESSMENT

As in the case of thermal and fast neutrons, activation through nuclear reactions can be used for high energy particle fluence assessment. Unfortunately the situation is complicated for two reasons:

(a) Formation of a radionuclide (whose activity would be proportional to the particle fluence) through several nuclear reactions initiated by different particles.

(b) Formation within a detector, together with the radionuclide of interest, of a large number of different radionuclei, whose radiations make it difficult to assess the activity of the radionuclide of interest.

These considerations make it clear that an unambiguous interpretation of the response of an activation detector for a particular nuclear reaction is possible only in the presence of a pure beam of monoenergetic high energy particles.

In the presence of a composite or unknown radiation field, the technique is of more restricted value. It may for example be necessary to study the responses of a group of detectors (rather than a single detector) in order to obtain a valid assessment of particle fluence and energy spectrum. An example of this method is given in section D.2.

C. In-beam Fluence Measurements

The most widespread use of activation detectors in high energy radiation measurements has been fluence assessment in a relative pure beam of primary particles and especially protons.

1. CHOICE OF THE TARGET AND MONITOR REACTIONS

Any material and reaction can in principle be used to monitor a pure beam of particles. The choice is more restricted if certain factors essential for an accurate measurement are considered.

The most important conditions are:

(a) The absolute cross section for the reaction should be known in the energy region under consideration.

(b) Cross section values must be sufficiently high to obtain adequate sensitivity from the detector.

(c) The half life of the radioisotope produced by the nuclear reaction considered should be longer than the irradiation time.

(d) It should be possible to assay the radioisotope of interest easily and without chemical separation. For this reason targets of low atomic number are usually the only suitable ones.

(e) The radiation emitted by the radioisotope must be convenient for detection.

(f) The reactions used should not be sensitive to secondary particles formed by interaction of the primary beam in the target and its surroundings. The reaction $^{12}C(p, pn)^{11}C$ is sensitive to neutrons which, through the reaction $^{12}C(n,2n)^{11}C$ produce the same isotope. The same consideration applies to the reaction $^{27}Al(p, 3pn)^{24}Na$, which is particularly sensitive to low energy neutrons through the interfering reaction $^{27}Al(n,\alpha)^{24}Na$ with a threshold of 6 MeV (see Table I).

(g) The materials constituting the activation detector must be very pure and in a form convenient for manipulation, thus excluding gases and brittle materials.

2. SOME REACTIONS USED TO MONITOR PURE PARTICLE BEAMS

Cumming (1963) has summarized the reactions most often used to monitor pure (mainly proton) beams of particles.

The reaction $^{12}C(p, pn)^{11}C$ is the primary standard since absolute values of the corresponding cross-sections are best known (Bruninx, 1964). This reaction has a threshold at 18·5 MeV (Panofsky and Phillips, 1948) and a relatively large cross section which, together with the short half life of the ^{11}C formed, renders it very sensitive for short time irradiations.

The positrons emitted by the ^{11}C isotope are easily counted with either a beta proportional counter or (through their annihilation radiation) by scintillation counting.

For longer irradiations the reaction $^{12}C(p, 3p3n)^{7}Be$ is sometimes

TABLE I

Some Detectors and Reactions used for Proton Detection by Activation

Detector	Isotope and Abundance %	Reaction	Reaction threshold MeV	Cross sections (mb) peak	Cross sections (mb) plateau	Radio-nucleus produced	Half life	Radiations and energy MeV	% emission per disintegration
Boron	^{11}B(81·3)	(p, n)	2	100		^{11}C	20·4 mm	β^+(0·97)	100
								γ (0·511)	200
Carbon	^{12}C(98·9)	(p, pn)	18·5			^{11}C			
Aluminium	^{27}Al(100)	$(p, 3pn)$	40		10	^{24}Na	15 hr	β^-(1·38)	100
								γ (1·37)	100
								(2·75)	100
Aluminium	^{27}Al(100)	$(p, 5p5n)$	40		7	^{18}F	109·9 mm	β^+(0·64)	97
								γ (0·511)	194
Carbon	^{12}C(98·9)	$(p, 3p3n)$	40	26		^{7}Be	53·0 d	γ (0·477)	10·3
Gold	^{197}Au(100)	$(p,$ spallation$)$	600		1	^{149}Tb	4·2 hr	α(3·95)	16

used but the weak emission of the 0·477 MeV γ-rays makes activity measurements difficult.

Activation of aluminium through the reactions ^{27}Al$(p, 3pn)^{24}$Na and ^{27}Al$(p, 5p5n)^{18}$F is widely used because of the ease of obtaining aluminium foils of different sizes and thicknesses and the convenient half lives of the reactions products.

These two reactions in aluminium have high and almost identical energy thresholds and their cross section values are roughly independent of energy.

The reaction (p, n) on ^{11}B, also yielding ^{11}C, can be used to assess proton fluences. The reaction is sensitive and has a low threshold. Some cross section values for energies higher than 150 MeV have recently been measured (Sklavenitis, 1966).

Very high energy protons (above 600 MeV) can be assessed through gold spallation yielding ^{149}Tb. This radionucleus decays with half life 4·2 hr emitting 3·95 MeV α-particles. The reaction is insensitive to most of the secondary nucleons which might interfere, but low cross section values and α-particle absorption in the gold foils renders its application difficult. As far as neutron measurements are concerned, the reactions ^{31}P$(n, p)^{31}$Si and ^{32}S$(n, p)^{32}$P, have been quite extensively studied from the threshold to about 20 MeV (Liskien and Paulsen, 1962). There is little information on cross section values at higher energies.

Phosphorus has a single stable isotope rendering neutron fluence assessment unambiguous. Sulphur has four stable isotopes (mass numbers 32, 33, 34, 36) which might produce competitive reactions. Fortunately the small abundances of ^{33}S(0·76%) and ^{36}S(0·014%) render these two isotopes insignificant in this connection but the reactions ^{34}S$(n, p2n)^{32}$P and ^{34}S$(p, 2pn)^{32}$P may sometimes be relevant when considering corrections.

Carbon-11 production through the reaction ^{12}C$(n, 2n)^{11}$C is a convenient method of neutron detection. The reaction threshold is close to 20 MeV and cross section values are known up to about 400 MeV (Warshaw et al., 1954; Organesian, 1953).

^{24}Na and ^{18}F produced in aluminium through the reactions ^{27}Al$(n, \alpha)^{24}$Na and ^{27}Al$(n, 4p6n)^{18}$F may also serve for neutron fluence assessment. Cross section variation in the low energy region is well know for the former but only a few measurements are available for the latter reaction—which is however commonly assumed, as a first approximation, to have the same threshold and mean plateau cross section values as the proton reaction ^{27}Al$(p, 5p5n)^{18}$F, also producing ^{18}F in aluminium.

Tables I and II summarize the properties of the above mentioned reactions; the corresponding cross section curves, presented in Figs

TABLE II

Some Detectors and Reactions used for Neutron Detection by Activation

Detector	Isotope and Abundance %	Reaction	Reaction threshold MeV	Cross sections (mb) peak	plateau	Radio-nucleus produced	Half life	Radiations and energy MeV	% emission per disintegration
Phosphorus	^{31}P(100)	(n, p)	1·6	150		^{31}Si	2·62 hr	β(1·48)	99·93
								γ(1·26)	0·07
Sulphur	^{32}S(95·02)	(n, p)	1·6	340		^{32}P	14·3 d	β^-(1·71)	100
Carbon	^{12}C(98·9)	$(n, 2n)$	20·2		20	^{11}C	20·4 mm	β^+(0·97)	100
								γ(0·511)	200
Aluminium	^{27}Al(100)	(n, α)	6	150		^{24}Na	15 hr	β^-(1·38)	100
								γ(1·37)	100
								(2·75)	100
Aluminium	^{27}Al(100)	$(n, 4p\ 6n)$	40	7		^{18}F	109·9	β^+(0·64)	97
								γ(0·511)	194

1 and 2, are based mainly on the work of Cumming (1963), Bruninx (1964), Charalambus *et al.* (1966) and Sklavenitis (1967).

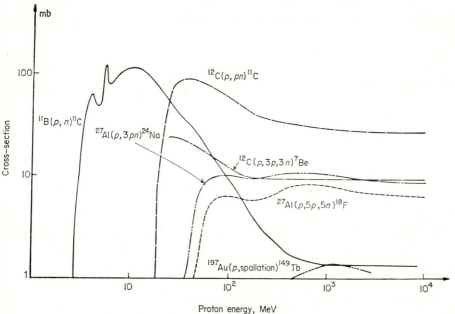

FIG. 1. Cross-section *versus* energy curves for some proton induced spallation reactions.

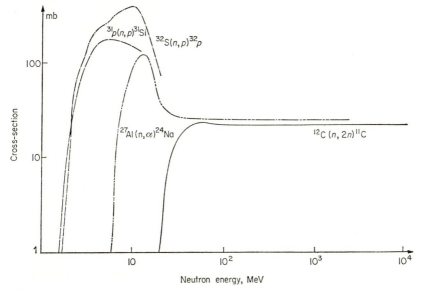

FIG. 2. Cross-section *versus* energy curves for some neutron induced spallation reactions.

3. DOSE CALCULATION

The knowledge of the particle fluence Φ of a particular group of particles allows the calculation of the absorbed dose D which is proportional to it.

$$D = A\Phi \quad (1)$$

where A is the absorbed dose per unit fluence for that group of particles.

For monoenergetic particle fluences, A is a constant for a given energy. If the fluence is described by a spectrum distribution $\phi(E)$ particles cm^{-2} MeV^{-1}, the mean value of A is

$$\bar{A} = \frac{\int A(E)\phi(E)\,dE}{\int \phi(E)\,dE}. \quad (2)$$

Thus the calculation of A presupposes a knowledge of the fluence spectral distribution, which is one part of the information sought through the use of activation techniques. In practice, plausible spectra are presumed—and verified by the concordance of responses of a group of activation detectors. The problem naturally becomes more complicated in the presence of particles of more than one kind, each with a different energy spectrum.

If Φ is inferred by the activity of a detector then

$$\Phi = a/\sigma \quad (3)$$

where a is proportional to the activity at the end of irradiation (Sklavenitis, 1967) and σ is the cross section of the reaction producing the activity at a definite energy. The average cross section $\bar{\sigma}$, is used when the fluence is not monoenergetic:

$$\bar{\sigma} = \frac{\int \sigma(E)\phi(E)\,dE}{\int \phi(E)\,dE}. \quad (4)$$

Finally the absorbed dose is given by the relation:

$$D = A\,a/\sigma. \quad (5)$$

Values of A have been calculated by the Monte Carlo method for protons of various energies (and in some cases for neutrons) incident on a tissue-equivalent phantom.

TABLE III

Values of A for Perpendicularly Incident Protons at the Surface of a Tissue Equivalent Phantom.

Proton energy MeV	100	200	400	600	1000
rad p^{-1} cm^{-2} × 10^{-8}	14·0	8·0	5·8	4·5	4·0

Table III gives calculated values of absorbed dose per unit fluence of perpendicularly incident protons at the surface of a tissue-equivalent phantom (Turner et al., 1964; Wright et al., 1966).

Values of A' corresponding to dose equivalent per unit particle fluence incident on a phantom have also been calculated by the same method.

D. Composite Radiation Fields

When dose measurements are performed inside targets bombarded by a beam of primary particles, or in the stray radiation fields around high energy accelerators, one is normally concerned with mixed radiation fields. In that case fluence values have to be accompanied by information on the nature of the particles as well as their energy spectra if radiation hazards are to be assessed. Uncertainty regarding energy spectra usually leads to serious errors in the estimation of fluence.

1. MEASUREMENTS IN MIXED RADIATION FIELDS

In order to evaluate absorbed doses received by individuals working near high-energy accelerators, activation detectors are currently used to classify particle fluences into broad groups to which an approximate value of $\overline{A}$ can be attributed.

The nuclear reactions through which activation is obtained may be selected from among those mentioned in section C.2, or others depending on the particles concerned, their energy range and irradiation conditions.

The Berkeley group (McCaslin et al., 1966) use the (n, p), (n, α) and $(n, 2n)$ reactions simultaneously on several detectors, including Mg, Al, Ni and Co to assess neutron fluences and define their spectra in the limited energy region of 2–30 MeV.

A cobalt disc, surrounded by a cadmium covered hydrogenous moderator serves through the reaction $^{59}Co(n, \gamma)^{60}Co$ to detect fast neutrons up to several MeV in energy (Smith, 1961). The 5·3 year half-life of ^{60}Co permits fluence estimations to be obtained from exposure times as long as one year.

Neutrons and protons, and incidentally π-mesons, with energies greater than 20 MeV can be easily detected, without distinction, by the $(n, 2n)$ and (p, pn) reactions on carbon (McCaslin, 1960). This detector is particularly sensitive if used in the form of a transparent phosphor-type plastic (McCaslin, 1960; Charalambus et al., 1966) which is assayed, after irradiation, by coupling directly to a photomultiplier. In this arrangement the weakest radiation field detectable is of the order of 1 particle cm^{-2} sec^{-1}. An important limitation to the use of carbon as a fluence detector for neutrons and protons is introduced by the ^{12}C$(\gamma,n)^{11}$C reaction in the presence of substantial high-energy photon fluences.

Terbium formation in gold, by spallation, can be used to measure particle fluence at energies greater than the threshold at about 600 MeV. In practice, this reaction in gold is not easily applicable because of its low sensitivity.

Another possibility is to use Hg as an activation detector, measuring the ^{149}Tb activity after chemical extraction (McCaslin, 1967) which gives a yield of about 60%. The amount of ^{149}Tb activity produced per gram of mercury is of the same order of magnitude as that produced in gold but better sensitivity can be obtained when the terbium is extracted from a large volume of mercury. A sensitivity of about 3×10^{-2} cpm per proton cm^{-2} sec^{-1} can be obtained with 500 gram mercury samples. Since a background count rate of 0·1 cpm is easily obtained with a methane-flow counter, a flux of a few protons cm^{-2} sec^{-1} is detectable by mercury activation. The technique is useful for neutrons also, since the reaction cross sections are about the same.

Aluminium activation is also used to assess proton and neutron fluences. ^{24}Na producing reactions are the more sensitive but, in a mixed radiation field, fluences obtained are quite approximate because of differences in reaction thresholds and cross section values. Moreover cross sections of the neutron-initiated ^{24}Na producing reaction are not accurately known above 20 MeV.

For ^{18}F formation in aluminium by protons and by neutrons a reaction cross section of 7 millibarns is assumed for both particles and a common reaction threshold of about 40 MeV. Under these assumptions particle fluences assessed through fluorine formation are more meaningful.

Fluence assessment through ^{7}Be formation in light elements offers the advantage (over ^{11}C) of allowing, because of the 53-day half life, the longer integration times useful for protection measurements.

It has been mentioned in section C·2, that poor counting efficiency is associated with the disintegration scheme of ^{7}Be. A separation process for extracting ^{7}Be from large volumes of liquid targets uses a series of filters (McCaslin et al., 1966), each extracting about 50% of the ^{7}Be.

The ^{7}Be is formed in carbon (in the form of benzene), in oxygen (in the form of water) and in nitrogen (in the form of liquid nitrogen). At present cross section values are known only for proton induced reactions in carbon. Assuming, somewhat arbitrarily, a 10 millibarns cross section for all reactions, sensitivity has been calculated to be of the order of 1 cpm in the 0·477 MeV γ peak per litre of liquid detector, per unit particle fluence rate. The background count rate of a conventional sodium iodide detector for the same counting conditions and γ energy region is reported to be 12 cpm (McCaslin et al., 1966). These methods give limited accuracy in the estimation of particle fluences. This is mainly due to the fact that assumptions have to be made on particle spectra and cross section values which are often incompletely known.

Nevertheless the activation analysis technique is often a convenient method of fluence measurement and the results offer sufficient accuracy for protection purposes.

2. DOSIMETRY WITHIN A PHANTOM

In order to estimate component fluences in a mixed beam, it is usually necessary to use a number of activation detectors simultaneously. When this is done, the reliability of assumed energy spectra for various components may be assessed by comparison with the corresponding activities in the detectors. However, an increase in the number of detectors increases the associated labour to a degree which often necessitates the use of a computer for the calculations.

In some cases a number of simlpifying assumptions are possible, leading to a quite satisfactory fluence analysis using a limited number of activation detectors.

An example (which is important in relation to the possibility of an irradiation accident in the vicinity of a high energy particle accelerator) is provided by recent work on fluence analysis in the interior of a tissue equivalent phantom irradiated by a beam of perpendicularly incident protons of energy a few GeV (Sklavenitis, 1967, 1968a).

The activation detectors used are boron, carbon, aluminium and sulphur in elemental form as pure as possible. For these elements, reaction thresholds and cross sections are known in considerable detail. The corresponding nuclear reactions and decay characteristics of reaction products are shown in Tables I and II.

For a tissue equivalent phantom (a cylinder 20 cm in diameter and 30 cm long) bombarded by a homogeneous beam of protons of energy 3 GeV and of diameter comparable of that of the phantom, the main assumptions are the following:

(a) π-meson fluence is much less than secondary proton fluence. No attempt has been made to distinguish the π-mesons as a different group.

(b) The induced activity per nucleus, for proton activation reactions involving primary protons varies with the depth x along the phantom axis, according to the relation:

$$(\sigma\Phi)_x = \Phi_o \sigma \, e^{-x/\lambda}. \tag{6}$$

Where Φ_o is the incident primary proton fluence, σ the corresponding activation cross section at the primary proton energy and λ the mean free path of the primary protons within the phantom material.

This relation allows reasonable accurate identification of that part of the response of the proton detectors attributable to primary protons.

With these assumptions the analysis attempted refers to three main groups of particles:

(1) Primary protons.

(2) Cascade and evaporation secondary protons resulting from spallation reactions, together with those produced by elastic collision of nucleons with hydrogen nuclei forming part of the phantom constituents. (For an exactly tissue equivalent material hydrogen should contribute 10·2% by weight).

(3) Secondary cascade or evaporation neutrons resulting from spallation reactions.

The fluence analysis is based on the fact, already mentioned in section D.1. that the reactions $^{27}Al(p, 3pn)^{24}Na$ and $^{27}Al(p, 5p5n)^{18}F$, have practically a common threshold at about 40 MeV and that their respective cross sections are independent of proton energy spectra.

At first, primary protons are monitored in front of the phantom entrance and at some distance from it in order to avoid secondary radiations. Φ_o is found in this way, for subsequent use in equation (6). Next, a number of neutron differential energy spectra, $\phi(E)$, have been constructed. These spectra are similar to those calculated after the cascade and evaporation processes as described in the references quoted in section B.1. They generally related to particles with degraded energy and are schematically represented by segments proportional to a power of energy. An example of such a spectrum is:

$$
\begin{array}{ll}
1\text{--}100\text{MeV} & \phi(E) = k_1 E^2 \\
100\text{--}250 \text{ MeV} & \phi(E) = k_2 E^{-1} \\
250\text{--}3000 \text{ MeV} & \phi(E) = k_3 E^{-2}.
\end{array}
$$

For each neutron spectrum considered, a mean cross section $\bar{\sigma}$ for the reaction $^{32}S(n, p)^{32}P$, is calculated according to equation (4), section C.3. From the ^{32}P activity of the sulphur detector and the mean cross section, a neutron fluence is calculated. This neutron fluence is used to calculate what portion of the ^{24}Na activity in the aluminium detector is attributable to neutrons (by the n, α reaction on ^{27}Al). The remaining ^{24}Na activity in the aluminium detector, attributable to protons of energy above 40 MeV, by the reaction $^{27}Al(p, 3pn)^{24}Na$, is then used to estimate the corresponding proton fluence.

A similar calculation is performed in relation to the ^{18}F activity in the aluminium detector, which may be produced by a neutron reaction ^{27}Al$(n, 4p6n)^{18}$F or by protons with energy above 40 MeV in the reaction ^{27}Al$(p, 5p5n)^{18}$F. The neutron spectrum which leads most nearly to the same estimate of proton fluence by these two routes is adopted for subsequent calculations.

When once the neutron spectrum has been established, a number of similar models of proton differential energy spectra are tested by the compatibility among the integral proton spectra calculated through the activity induced by the reaction ^{11}B$(p, n)^{11}$C in boron and the residual activities induced by the reaction ^{12}C$(p, pn)^{11}$C in carbon and the reactions ^{24}Al$(p, 3pn)^{24}$Na and $^{24}(p, 5p5n)^{18}$F in aluminium.

The assessment of compatibility can be further illustrated as follows. A differential proton spectrum $\phi(E)$ is assumed for the incident beam. A proton fluence Φ_1, corresponding to energies above 2 MeV, is calculated from the ^{11}C activity produced in the reaction ^{11}B$(p, n)^{11}$C with a threshold of 2 MeV. Similarly a proton fluence Φ_2 is calculated by reference to the reaction ^{12}C$(p, pn)^{11}$C which has a threshold of 20 MeV. The plausibility of the assumed proton spectrum $\phi(E)$ depends on the extent to which the ratio Φ_1/Φ_2 conforms to the theoretical value calculated from the assumed spectrum—that is

$$\int_{2\text{MeV}}^{E\max} (E)\, dE \Big/ \int_{20\text{MeV}}^{E\max} (E)\, dE.$$

In other words if the integral spectrum $\int_{E\min}^{E\max} \phi(E)\, dE$ is plotted against energy, the proton fluences calculated from each proton spallation reaction should be correctly situated on the curve at the points corresponding to each reaction threshold.

These relatively simple techniques allow the calculation of proton and neutron fluences at each point of the phantom.

Figure 3 shows proton and neutron fluences with energies greater than the lower threshold of the reaction used—(p, n) on boron and (n, p) on sulphur—as a function of depth in the phantom along its principal axis of symmetry. Study of these fluences, along with the corresponding differential energy spectra, suggest some general observations:

(i) The secondary particle fluence increases with depth in the phantom.

(ii) Proton differential energy spectra become harder (richer in high energy protons) with depth in the phantom.

(iii) Neutron differential energy spectra do not change very much with depth in the phantom.

The same fluence analysis technique can be applied to any material target bombarded with high energy protons, such as concrete shielding blocks or accelerator magnets.

The use of these techniques in a stray radiation field is difficult for two reasons. First, particle fluences are often not high enough to produce

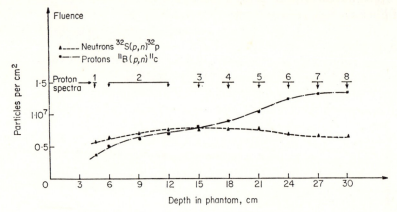

Fig. 3. Secondary proton and neutron fluences with energies greater than the corresponding reaction thresholds as a function of phantom depth. (Numbers 1–8 correspond to different proton spectra.)

accurately measurable activity in detectors using the reactions mentioned in the preceding paragraphs, thus necessitating a search for other reactions with longer lived products. Second, the general form of the particle spectra constituting the stray radiation are often unknown.

Detectors and reactions other than those already mentioned must be sought when it is required to deal with significant fluences of π-mesons (produced by bombardment with high energy protons) or X- and γ-rays from electron accelerators (section B.1).

E. Calculation of the Absorbed Dose

In order to evaluate the absorbed dose in tissue due to particle fluences, values of $\overline{A}$ have to be calculated for each fluence measured.

For a tissue equivalent phantom, considering the mechanisms of energy delivery by high energy particles, it has been assumed (Sklavenitis, 1967) that absorbed dose is mainly delivered through proton ionization.

In that case $\overline{A}_p$ values, valid only for protons, need to be established; they are calculated according to equation (2) of section C.3, using knowledge of the differential spectrum and of the stopping power of protons in tissue as a function of energy.

Thus the absorbed dose D is taken, as a first approximation to be

$$D = \Phi_p \bar{A}_p \tag{7}$$

Absorbed doses calculated in this way are shown, as a function of depth of phantom, in Fig. 4. For comparison, absorbed doses measured with a

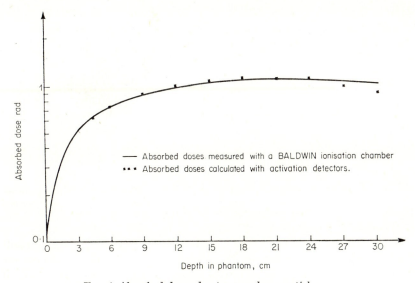

Fig. 4. Absorbed doses due to secondary particles.

Baldwin ionization chamber are also plotted (solid curve). The two sets of doses agree (except for the last point) with a precision better than 5%.

Dose equivalents (and subsequently quality factors) could also be calculated in the same way if variation of the dose equivalent per unit particle fluence as a function of energy were known for protons and neutrons. Such information being at present meagre, only very rough estimation of dose equivalent can be made. Accuracy is expected to increase as better dose equivalent data for high energy particles are accumulated.

F. Possibility of Use of Activation Detectors as Particle Dosimeters

In section C.3 the relation between absorbed dose, detector activity and reaction cross section has been established (equation 5).

Taking into consideration equations (2) and (4), the relation (5) is transformed to

$$D = \left[\int A(E)\phi(E) \, dE \bigg/ \int \sigma(E)\phi(E) \, dE \right] a. \tag{8}$$

If there exists between the function $A(E)$, representing dose variation as a function of energy, and the cross section function $\sigma(E)$ a relationship of the form

$$A(E) = k^{-1}\sigma(E) \tag{9}$$

where k (rad^{-1}) is a constant independent of the particle energy spectrum, then equation (8) can be written:

$$D = k^{-1} a. \tag{10}$$

In this way a simple measurement of the activity of a detector would suffice to assess the dose due to the particles whose energy is greater than the threshold of the reaction which produced the activity.

In the case of a tissue equivalent phantom irradiated by high energy protons, the dose is produced mainly through proton ionization (section D.2). The ideal detector for measuring the absorbed dose would be the one for which the ratio, $\sigma(E) \big/ \left(\dfrac{dE}{dx} \right)$, of the cross section of a nuclear reaction initiated by protons to the stopping power of tissue for protons, is a constant independent of energy.

Considering the shapes of the stopping power *versus* energy curve and the variation of cross section with proton energy, it follows that the most promising approach is the use of reactions such as (p, n) or (p, pn), giving only a small number of secondary nucleons (see also section B.2). For these reactions, the shape of the cross section curve is generally favourable and the relatively low threshold allows a large part of the proton spectrum to be taken into account in the calculation of absorbed dose. The reaction (p, pn) on carbon partially fulfils the requirements, but a (p, n) reaction would be more appropriate.

Figure 5 gives the variation of k as a function of energy for the reaction $^{12}C(p, pn)^{11}C$. It is evident that for protons with energy larger than about 500 MeV, k is effectively constant and its value is close to 8×10^{-19} dis/nucleus rad. This conclusion has been experimentally verified by measurement in a 3 GeV proton beam.

For proton spectra containing a lower energy component, such as those met inside tissue equivalent phantoms, the value of k diminishes by about 30%. As this value varies slowly with spectral form, it is possible, if the general shape of the spectrum is known, to evaluate also the absorbed dose due to protons above 20 MeV with a precision of the order of 10%.

It is also possible in principle to measure the dose equivalent directly if activation detectors with adequate cross section curves can be found; the various possibilities for absorbed dose or dose equivalent measurement have not yet been adequately studied. It is of course possible

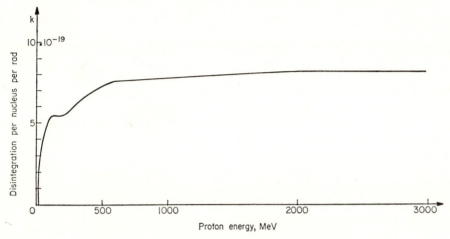

Fig. 5. Variation of k with proton energy for the (p, pn) reaction in carbon.

to consider a detector consisting of a mixture of elements in order to obtain a particular shape of cross section curve for a specified reaction. A mixture of boron and carbon could, for example, be used for the ^{11}C activation reactions.

G. Conclusions

Activation detectors can be used with high energy radiations to measure fluences in pure particle beams, to obtain a rough estimate of particle fluences in mixed radiation fields and, in certain cases, to analyse such fields into the main radiation components and their energy spectra. The first two operations are easy to carry out, but the third is more cumbersome and its complexity increases as greater refinement is sought.

The main advantages of activation detectors are as follows:
(a) Small volume.
(b) No possibility of saturation.
(c) Reliability over a wide range of energy.
(d) Absence of electronic equipment at the point of measurement.

(e) Possibility of many simultaneous measurements, for example to resolve a complex beam or field of radiation and particles.

(f) Possibility of estimating absorbed dose (through knowledge of particle fluences) and of assessing radiation hazard to individuals.

The main disadvantages are:

(a) Lack of sensitivity for weak radiation fields.

(b) Lack of precision because of incomplete knowledge of high energy reaction cross sections and uncertainty in defining energy spectra.

Improvement in the effectiveness of activation techniques can be expected in the following directions:

(a) Compliation of better data on reaction cross sections (especially for neutron reactions) and an energy spectra of secondary particles.

(b) Use of more selective counting techniques (as with germanium detectors) will allow a greater variety of reactions to be used for fluence estimation.

(c) Further availability of computers for data analysis will allow simultaneous consideration of a greater number of reactions.

(d) More accurate estimation of high energy radiation hazards will come with better knowledge of the relationships linking dose equivalent with particle fluence and energy.

References

Alsmiller, R., Leimdorfer, M. and Barish, J. (1967). Analytic representation of nonelastic cross section and particle-emission spectra from nucleon-nucleus collisions in the energy range 25 to 400 MeV. *ORNL* 4046.

Bertini, H. (1963a). Low energy intranuclear cascade calculations. Monte Carlo calculations on intranuclear cascades. *ORNL* 3383. *Phys. Rev.* **131**, 1801.

Bertini, H. (1963b). Parametric study of calculated cascade and evaporation reactions for 25–400 MeV nucleons incident on complex nuclei. *ORNL* 3499. **2**, 31.

Bertini, H. (1965). Results from low-energy intranuclear cascade calculations. *ORNL*–TM–1225.

Bruninx, E. (1961). High-energy nuclear reaction cross sections: I. *CERN* 61–1.

Bruninx, E. (1962). High-energy nuclear reaction cross sections: II. *CERN* 62–9.

Bruninx, E. (1964). High-energy nuclear reaction cross sections: III. *CERN* 64–17.

Charalambus, S., Dutrannois, J. and Goebel, K. (1966). Particle flux measurements with activation detectors *CERN*/DI/HP 90.

Cumming, J. (1963). Monitor reactions for high energy proton beams. *A. Rev. nucl. Sci.* **13**, 261.

Dostrovsky, I., Rabinowitz, P. and Bivins, R. (1958). Monte Carlo calculations of high-energy nuclear interactions. I. Systematics of nuclear evaporation. *Phys. Rev.* **111**, 1659.

Hess, W., Patterson, H., Wallace, R. and Chupp, E. (1959). Cosmic ray neutron energy spectrum. *Phys. Rev.* **116**, 445.

Hudis, J. and Miller, J. (1959). High energy nuclear reactions. *A. Rev. nucl. Sci.* **9**, 159.
Liskien, H. and Paulsen, A. (1962). Compilation of cross sections for some neutron induced threshold reactions. *EUR* 119 e.
McCaslin, J. (1960). A high energy neutron flux detector. *Hlth Phys* **2**, 399.
McCaslin, J., Patterson, W., Smith, A. and Stephens, L. (1966). Some recent developments in the technique for monitoring high-energy accelerator radiation. *UCRL*–16769.
McCaslin, J. and Stephens, L. (1967). High sensitivity neutron and proton flux detector with a practical threshold near 600 MeV, using Hg (spallation) ^{149}Tb. *UCLR*–17505.
Metropolis, N., Bivins, R., Storm, M., Miller, J., Friedlander, G. and Turkevich, A. (1958). Monte Carlo calculations on intranuclear cascades: II high energy studies and pion processes. *Phys. Rev.* **110**, 204.
Moyer, B. (1961). Evaluation of shielding requirements for the improved Bevatron. Rapport americain *UCRL* 9769.
O'Barrell, M. (1964). The calculation of proton penetration and dose rates, In Second Symposium on Protection against Radiation in Space. *NASA* SP–71, 493.
Organesian, K. (1953). Report of the Institute of Technical Nuclear Problems Acad. Sci. USSR quoted by Baranov, P. M. and Goldanski, V. (1955). High threshold scintillation neutron detector, Soviet Phys. JETP Translation, 1, 576.
Panofsky, W. and Phillips, R. (1948). Evidence for a (p, d) reaction in carbon. *Phys. Rev.* **74**, 1732.
Poskanzer, A., Cumming, J., Friedlander, G., Hudis, J. and Kaufman, S. (1961). ^{12}C (π-, π-n)^{11}C cross section at 1·0 BeV. *Bull. Am. phys. Soc.* **6**, 38.
Reeder, P. L. (1962). Nuclear reactions induced by pions and protons. *UCRL*–10531. CTID 4500.
Rossi, B. (1956). "High-Energy Particles". Prentice-Hall.
Sklavenitis, L. (1966). Sections efficaces de la reaction (p, n) sur les noyaux du bore-11 entre 150 MeV et 3 GeV. *C.r. hebd Séanc Acad. Sci., Paris* **263**B 833.
Sklavenitis, L. (1967). Sur la mesure et l'analyse des rayonnements de haute energie par detecteurs a activation—Application a la dosimetric. *CEA*–R 3376.
Sklavenitis, L. (1968a). Possibilités d'utilisation des detecteurs a activation comme dosimetres de particules de haute energie. *Nucl. Instr. Meth.* **58**, 176.
Sklavenitis, L. (1968b). Dosimetrie par detecteurs a activation dans des fantomes soumis a l'irradiation de protons de haute energie. *Nucl. Instr. Meth.* **59**, 73.
Smith, A. (1961). A cobalt neutron flux integrator. *Hlth Phys.* **7**, 40.
Turner, J. E., Zerby, C. D., Woodyard, R. L., Wright, H. A., Kinnen, W. E., Snyder, W. A. and Neufeld, J. (1964). Calculation of radiation dose from protons to 400 MeV. *Hlth. Phys.* **10**, 783.
Wallace, R. and Sondhaus, C. (1962). Techniques used in shielding calculations for high-energy accelerators: applications to Space Shielding. *TID* 7652, **2**, paper F–4.
Warshaw, S., Swanson, R. and Rosenfeld, A. (1954). Cross sections for the reactions ^{12}C(p, pn) (n, 2n)^{11}C. *Phys. Rev.* **95**, 649.
Wright, H., Branstetter, E., Neufeld, J., Turner, J. and Snyder, W. (1966). Calculation of radiation dose due to high-energy protons. Rapport presente au ler congres de l'IRPA–Sept. Rome.

γ-RAY SPECTROSCOPY BY MEANS OF GERMANIUM LITHIUM-DRIFTED DETECTORS IN ACTIVATION ANALYSIS

F. GIRARDI

and

G. GUZZI

Euratom, Ispra, Italy

A. Introduction	137
B. Theory and Preparation of Ge-Li Drifted Detectors	139
C. Commercial Detectors, Detector Mountings and Cryostats . . .	143
D. Electronic Circuits	144
1. Preamplifiers, amplifiers and analysers	145
E. γ-Ray Spectroscopy by means of Germanium Semi-Conductor Detectors	146
1. Determination of γ-ray energies—identification of γ-ray emitters .	146
2. Determination of counting rates and absolute γ-ray emission rates .	148
F. Applications to Activation Analysis	152
1. Analytical sensitivity	152
2. Fields of application	155
References	160

A. Introduction

Although semiconductor detectors have been used for only a few years, their importance as particle or γ-ray detectors has grown considerably and their development is still expanding.

Among the large variety of semiconductor detectors developed for different applications in nuclear physics, the Ge(Li) drifted detector is the one almost exclusively used in activation analysis as a high-resolution γ-ray detector. We shall therefore deal only with these detectors.

There has been a real advance towards higher resolutions in recent years, but the ultimate line-width attainable is apparently still far from being reached (Heath et al., 1966). At the same time the efforts to commercialize the detectors have resulted in new γ-spectrometers which can be bought at reasonable prices and used in the routine conditions of the analytical laboratories without serious difficulties. In this chapter we shall try to summarize the state of the field from the particular point of view of the activation analyst.

For this reason theory and methods of preparation have only been briefly summarized, while more attention has been given to the problems associated with the choice of the whole spectrometry system, its correct set-up, calibration and use. A short review of applications is also given to show the kind of problems where high resolution γ-ray spectrometry can be advantageous.

The review is not at all exhaustive, and only shows a few selected applications in various areas. More recent information on applications can be found in the Proceedings of the Third International Conference on Modern Trends in Activation Analysis, held in Gaithersburg, 7–11 October, 1968.

Readers who would like to study more closely the fundamental aspects of semiconductor detectors should consult the books by Taylor (1963), Dearnaley and Northrop (1964), and by Bertolini and Coche (1968). Only this last deals extensively with Ge(Li) drifted detectors.

The proceedings of a number of conferences held in recent years are also excellent sources of information. Among these have been:

(a) The 10th Summer Meeting of Nuclear Physicists held in Herceg-Novi, August–September 1965 (Nuclear Instruments and Methods, 1966, **43**, 1).

(b) The 10th Scintillation and Semiconductor Counters Symposium held in Washington in March 1966 (IEEE Transactions on Nuclear Science, 1966, NS, *13*, 3). A summary of the highlights of the symposium was also published in Nucleonics (Glos, 1966).

(c) The IAEA Panel on Lithium Drifted Germanium Detectors, Vienna, June, 1966. (I.A.E.A., 1967).

(d) The 11th Scintillation and Semiconductor Counters Symposium, Washington, February 1968. (IEEE Transactions on Nuclear Science, NS, **15**, 3).

The meeting on Special Techniques and Materials for Semiconductor Detectors, held in Ispra, October, 1968. The proceedings will be published as a Euratom Report.

Finally, a detailed description of the most recent developments on semiconductor detectors and associated electronics has been presented by R. L. Heath at the 1968 International Conference on Modern Trends in Activation Analysis.

B. Theory and Preparation of Ge-Li Drifted Detectors

A detailed treatment of the theory of semiconductor detectors and their preparation methods would be outside the scope of this work.* Only a descriptive theory will be given to clarify the functioning of this kind of detector and to justify the special techniques and electronic accessories which they require. Ge-Li drifted detectors are essentially solid state versions of ionization chambers. If a single crystal of germanium (which is a tetravalent element) is grown in presence of a pentavalent element such as P, As, Sb, the final crystal will contain these doping elements in some lattice positions normally occupied by germanium atoms. As the Ge atoms are bound together by covalent bonds, the doping elements cause the presence of extra electrons. The new bond is rather weak due to the high dielectric constant of the medium (~ 16 for germanium) and thermal excitation, even at room temperature, is sufficient to free the electron and leave a fixed positive charge in the lattice.

The impurity acts therefore as an electron donor and n-type germanium is thus created. When subjected to an electric field, the crystal will conduct electricity by movement of the donor's free electrons. Conversely, if the crystal is grown in presence of trivalent elements (B, Ga, In), the addition of these elements produces holes in the electron conduction band (which behave as positive charges) and fixed negative charge centres. A p-type germanium semiconductor is thus created. When subjected to an electric field, the crystal will conduct electricity by movement of the holes, which can be imagined as a retrograde motion of electrons from filled positions in the valence band, to fill the holes.

When a readily ionizable metal such as lithium is diffused interstitially through one surface of a p-type germanium, the crystal region in which lithium is present will behave as an n-type germanium, as the number of free electrons from Li will by far exceed the number of free holes from the trivalent doping element.

At the interface of the n- and p- regions a p-n junction is formed; the electrons of lithium will compensate the holes created by the trivalent impurity and no excess of holes or electrons is present. A depletion layer is thus formed in which the only kind of electrical conduction possible is that due to thermal excitation of the electrons. This conduction mode (intrinsic conduction) is important in germanium at

---

* For a comprehensive account, reference should be made to the Proceedings of the 10th Summer Meeting of Nuclear Physicists, held at Herceg-Novi in 1965 Nuclear Instruments and Methods (1966), **43**, 1.

room temperature or above, but it disappears when the detector is cooled to liquid nitrogen temperature.

If a negative voltage is applied to the p-side of the crystal (reverse bias) at a temperature of 40–60°C, the lithium ions, which have a high mobility in germanium crystals at this temperature, drift slowly towards the p-side, and compensate in their migration the negatively charged lattice ions of the doping acceptors, by forming neutral electric dipoles. As long as the compensation proceeds, new lithium is attracted from the n-side so extending the depth of the depleted region. The intrinsic conduction of the compensated layer under the applied bias introduces a competing process to the lithium ion drift:

(1) the electrons which are collected at the n-region (positive electrode) tend to compensate the positive charge of the excess lithium ions, and therefore lower their mobility under the drift potential,

(2) the positive charges which are collected at the p-region tend to replace the lithium ions in the compensation of the negative fixed centres of the p-region.

As the intrinsic conduction increases, of course, with the depth of the compensated region, the competition of the charge carrier collection will also increase up to the point at which the Li-drift process will stop.

The depth of the intrinsic region is given in millimetres by the formula:

$$\omega = \sqrt{2\mu \, Vt} \qquad (1)$$

where V is the drift reverse bias, t is the drift time and μ is the lithium ions mobility in germanium, which is related to the diffusion coefficient D of the Einstein relation:

$$D = \frac{KT}{e}\mu \qquad (2)$$

where K is the Boltzmann constant, T is the absolute temperature, and e is the electronic charge. In practice it is difficult to obtain compensated layers deeper than 10 mm.

When the crystal with the applied bias is cooled to liquid nitrogen temperature the intrinsic conduction is stopped. The crystal is now ready to operate as a detector.

If a γ-ray strikes the intrinsic region, the ionization produced by the absorption process generates electron-hole pairs which are rapidly collected at the electrodes, giving rise to a voltage pulse proportional to the number of pairs formed and therefore to the energy of the γ-ray quantum, assuming that the electrons released in photoelectric, Compton or pair production processes are completely stopped in the intrinsic region. The greater the volume of the intrinsic region, the greater the detection efficiency. Larger intrinsic volumes may be obtained by coaxial drifting (Tavendale, 1966; Ewan et al., 1966).

The detectors must always be kept at a temperature below −80°C. Two phenomena can occur to damage the detector. First, because of the weakness of the pairing bond, if the temperature is sufficiently high lithium can move away from the acceptor centre. Second, because in the n-region lithium is present at an abnormally high concentration, precipitation can be encouraged by the mobility of the lithium ions and the presence of structural defects in the crystal. A few hours at room temperature are sufficient to produce appreciable changes with loss of detector performance.

The important advantage of the p-i-n diode over the scintillation detector is that the energy necessary to create an electron-hole pair is only about 3 eV, while an energy expenditure of at least 300 eV in a NaI(Tl) crystal is needed to secure the release of one photoelectron from one cathode of one associated photomultiplier. Statistical fluctuations in the pulse production process are much smaller for a solid-state detector; the resolution of semi-conductor detectors is therefore, inherently higher.

The theoretical resolution of a detector is:

$$R = 2.355\sqrt{FE\epsilon} \qquad (3)$$

where E is the energy of the γ-ray in keV, ϵ is the average energy in eV necessary to create an electron-hole pair and F is the Fano factor which is related to the fractional amount of total energy absorbed in the production of electron-hole pairs. Most recent values of ϵ for germanium range from 2·94 to 2·98 eV at 77°K. The exact value of the Fano factor is not yet known. Recent determinations (Mann et al., 1966) quote 0·13–0·16, although a lower value is possible. Figure 1 shows resolution *versus* γ-ray energy based on equation 3 (curve 1), and experimental resolution for a 2 cm³ planar detector (Heath et al., 1966) (curve 2), and a 47 cm³ coaxial detector (curve 3). The resolution of a 3×3 in. NaI (Tl) scintillator is shown for comparison (curve 4).

The experimental resolution is considerably inferior to that expected from theoretical considerations, because of additional noise from detector leakage and from auxiliary electronic currents.

The construction of the detectors proceeds through the following steps:

(1) *Preparation of the Ge(Li) crystals:* the Ge ingots must be especially chosen as the success obtained in the preparation depends on the quality of the starting material. P-type germanium with a high resistivity (>5 Ω cm) doped with In or Ga is used.

A low dislocation density (less than 2500 EPD/cm²† with no lineage) and a high carrier lifetime (>50–100 μsec) are necessary.

The ingots are cut in the desired shapes, lapped and etched by various procedures.

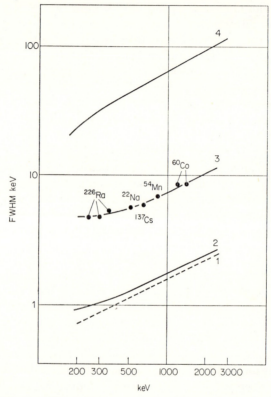

Fig. 1. Resolution against γ-ray energy. Curve 1: theoretical resolution obtained by relation (3) with a Fano factor of 0·15. Curve 2: experimental resolution of a 2 cm³ planar detector (Heath *et al.*, 1966). Curve 3: experimental resolution of a 47 cm³ coaxial detector (Girardi *et al.*, 1967). Curve 4: resolution of a 3×3 in. NaI(Tl) crystal, reported for comparison. (From Girardi *et al.*, 1967).

(2) *Li diffusion:* Li metal is deposited on the etched crystal, e.g. by vacuum evaporation, and then diffused into the crystal by heating, at about 400°C, for a few minutes.

(3) *Drift*: this is carried out by applying a bias (up to about 1000 volts) to the crystal at a controlled temperature. An easy way of controlling the temperature is to apply the bias to the crystal immersed

† Etch pit density.

in a low-temperature boiling organic solvent. Liquids with a zero dipole moment are preferred (Cappellani et al., 1967). The heat generated by the reverse current (up to some hundreds of mA) is removed by the boiling of the liquid. The drifting process, which may last for several months, is usually completed by applying a reverse bias of 500–1000 V for a short time at a temperature of $-10°C$.

(4) *Mounting of the detector and final clean-up drift*: the detector is mounted in a vacuum chamber and a final clean-up drift (at low temperature) is performed. The detector is cooled to liquid nitrogen temperature and it is ready for operation.

C. Commercial Detectors, Detector Mountings and Cryostats

Ge(Li) detectors can be divided into two main groups:

First, *planar detectors*: these are generally cylindrical in shape with surfaces up to 6 cm^2 and depletion depth up to 10 mm. The volume (6 cm^3) is perhaps too small for general purpose applications in activation analysis, but the resolution is good.

Second, *coaxial detectors:* coaxial detectors are prepared by diffusing lithium on all surfaces of the detectors except one, and drifting it towards the inside of the detector. Coaxial cylindrical detectors in which lithium is diffused from the side surface and drifted towards the axis of a cylinder are also available.

The technique of coaxial drifting makes it possible to prepare bigger detectors, although their larger self-capacitance and imperfect charge collection make the resolution somewhat poorer. They are therefore particularly interesting for activation analysis, as higher efficiency largely compensates for the lower resolution in most applications. Coaxial detectors up to 60 cm^3 are produced commercially.

Two kinds of detector mounting are available:

First, *encapsulated detectors* are hermetically sealed in an aluminium or mild steel capsule. The advantage is that encapsulated detectors can be shipped and stored at dry-ice temperature. Planar detectors, and coaxial detectors up to 30 cm^3 are available in encapsulated form.

Second, *non-encapsulated detectors* must always be kept in vacuum at low temperature. The whole detector-chamber-cryostat assembly is therefore shipped (Fig. 2). Non-encapsulated detectors are available in many shapes and dimensions.

In Fig. 2 types A and B are based on standard Dewar flasks. The detectors are cooled by a copper rod immersed in the liquid nitrogen (dipstick cryostat). Twenty-five litres Dewar flasks are generally used, and can last 10–15 days before refilling.

In types C and D the Dewar flask is generally a 10 litre tank, which can last 5–6 days before refilling.

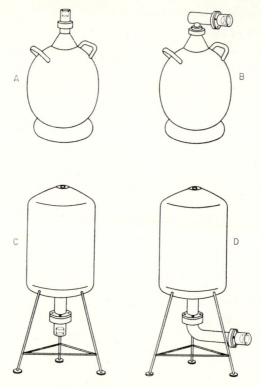

FIG. 2. Schematic representation of commercial detector-chamber-cryostat assemblies.

All of the detector chambers are designed to minimize scattering; the front window can be as thin as 0·3–0·5 mm. Vacuum (10^{-6}–10^{-7} Torr) is generally maintained by a small pump (of capacity 1 litre per sec).

D. Electronic Circuits

The absorption of a γ-ray quantum results in the production of a number of electron-hole pairs. The subsequent amplification and discrimination processes introduce a loss of resolution which is inherently different to that associated with a scintillation detector. There the photoelectron emission from the cathode of the phototube is multiplied 10^6–10^8 times by the dynode chain, practically without noise addition. The output pulse of the photomultiplier is therefore rather large (up to a few volts) and the fluctuating contribution of the background noise can often be neglected. The resolution of a scintillation γ-ray spectrometer depends almost entirely on the characteristics of the detector and depends very little on the experimental conditions, provided that unduly high counting rates are avoided.

In Ge(Li) detectors the collection of the charge carriers gives a voltage pulse of amplitude less than a mV, which must be considerably amplified. For a given charge collected, the amplitude of the voltage pulse is inversely proportional to the capacitance of the detector, which depends on the surface area of the detector and depth of the intrinsic region; typical values are less than 10 pF for a planar detector and 50 pF for a 50 cm³ coaxial detector.

Allowance must be made for the capacitance of the connection to one preamplifier. This effect is minimized by mounting the preamplifier as close as possible to the detector; the additional capacitance of the connection is usually 3–5 pF.

1. PREAMPLIFIERS, AMPLIFIERS AND ANALYSERS

The noise characteristics of the preamplifier are important in reducing loss of resolution associated with the electronic circuits. This loss is usually expressed by two terms (in keV), one independent of the capacitance of the detector, and one proportional to it.

Commercial vacuum-tube charge-sensitive preamplifiers give a resolution R which is of the order of 1·5 keV fwhm† at zero external capacitance while the contribution of the detector capacitance is of the order of 0·1 keV/pF. Charge sensitive preamplifiers based on the use of field effect transistors (FET) at room temperature give a resolution which is about 30–50% better. If the input FET is cooled to the optimum temperature (100–130°K) a further gain of 30–50% can be obtained. Low noise preamplifiers are still being improved in design and performance.

Amplifiers for semiconductor detectors differ from the conventional amplifiers for scintillation spectroscopy by their greater flexibility in pulse shaping. The pulse from the preamplifier can be differentiated, integrated and then differentiated again with different time constants to optimize the signal-to-noise ratio and minimize the distortion associated with pulse pile-up at high counting rates.

† Full width at half maximum.

The high resolution of the Ge(Li) detector requires the use of a 4000-channel pulse height analyser if the whole energy range up to 2–3 MeV is to be scanned without loss of resolution due to the finite width of the channels. When only a limited portion of the spectrum is to be examined, a 400- or 512-channel analyser, in conjunction with a biased amplifier, will be adequate.

Although large multichannel analysers are of course preferable, their price is too high for many analytical laboratories. The read-out and handling of the output data from such analysers also constitute a problem whose best solution seems to be a direct computer coupling, which further increases the price of the installation.

E. γ-Ray Spectroscopy by means of Germanium Semi-Conductor Detectors

1. DETERMINATION OF γ-RAY ENERGIES—IDENTIFICATION OF γ-RAY EMITTERS

A correct identification of radioisotopes in complex mixtures by accurately measuring the energy of the prominent peaks is the goal of qualitative analysis, and the advantages of semiconductor detectors in this respect are evident.

In good installations and with accurate measurements in conditions which minimize drift, an accuracy of about 0·1 keV in the assessment of γ-ray energies can be reached (Heath et al., 1966; White and Groves, 1967). In our own installation, with an 11 cm^3 detector the precision reached in repetitive counting of sources in condition of minimum drift was about 0·15–0·30 keV in the range 300–1400 keV. The energy of the ^{137}Cs peak (determined by assuming that the spectrometer calibration was linear between two calibration points at 511 and 1114 keV) was 0·5 keV lower than the accepted value of 661·595 (Black and Heath, 1967). By extrapolating the linear calibration up to 1368 keV the peak of ^{24}Na (1368·526 was found 1·4 keV lower, thus showing either non-linear calibration or inaccurate calibration points or both (Girardi et al., 1967).

In qualitative analysis such a high precision is difficult to reach and is not always necessary, since peaks with similar energies from radioisotopes formed by (n, γ) reactions usually differ by more than a few keV.

Linearity can be established by making a calibration curve with weak sources of well known energy. Table I lists a few radioisotopes whose γ-ray energies have been very precisely established with semiconductor detectors and which are ideally suited for calibration purposes (Black and Heath, 1967).

The drift behaviour of the spectrometer installation should be very carefully controlled. Drifts from both ambient temperature excursion and variation of counting rates have been found in our own installation (Girardi et al., 1967). The first reached 10 keV in 24 hr in a non-thermostatically controlled environment, the second reached 4–5 keV, when passing from weak ^{137}Cs sources to sources with a total counting rate of 5×10^5 cpm (50% dead time in our analyser).

TABLE I

Isotope	Energy (keV)
^{241}Am	59·568 ± 0·017
^{153}Gd	97·43 ± 0·02
	103·18 ± 0·02
^{177}Lu	112·952 ± 0·003
	208·359 ± 0·01
^{203}Hg	411·795 ± 0·009
^{198}Au	279·16 ± 0·02
Annihilation	511·006 ± 0·002
^{207}Bi	569·62 ± 0·06
^{137}Cs	661·595 ± 0·076
^{60}Co	1173·226 ± 0·040
	1332·483 ± 0·046
^{24}Na	1368·526 ± 0·044

The magnitude of both effects, although much smaller than the corresponding effects in scintillators are sufficiently great to cause concern.

The exact location of the photopeak axis can be difficult if the channel width is too large, if the counting statistics are poor or if a high background is present. A useful procedure is to calculate the width of the peak at a prefixed fractional height and assume that the peak axis passes through the middle (Black and Heath, 1967). In our routine computer runs we subtract the underlying background by linear interpolation between the two minima and calculate the axis F of the corrected peak by means of the equation:

$$F = \sum_{K_1}^{K_2} KC(K) / \sum_{K_1}^{K_2} C(K)$$

where the limit channels K_1 and K_2 are chosen so that the channel contents $C(K)$ between K_1 and K_2 are all larger than half the channel content of the maximum.

Once the energy of the experimental peaks has been assigned, the problem of correctly identifying the γ-emitters should be a minor one if accurate γ-ray energy data are available. In a recent work by Dams and Adams (1968), the γ-ray energies of the most prominent peaks of all isotopes produced by neutron capture are presented in a handy form for routine use. As a help in correct identification, a list of peaks from radioisotopes formed by (n, γ) reaction, in order of increasing γ-ray energy has been prepared by means of a computer programme (Guzzi et al., 1966). The list also gives associated peaks and their relative abundance for the particular detector and geometrical conditions used. A new list can be rapidly prepared whenever a fresh detector is obtained, by running the programme with the appropriate efficiency curve.

2. DETERMINATION OF COUNTING RATES AND ABSOLUTE γ-RAY EMISSION RATES

I. *Photopeak counting rates*

The accurate and simple measurement of photopcak areas has long been a problem in NaI(Tl) scintillation spectrometry. Various solutions, both mathematical and graphical, have been given (Schulze, 1962).

Most procedures require the estimation of the background underlying the peak. Linear interpolation of the background on both sides of the peak is the simplest method and is satisfactory provided that the background shapes in the peak neighbourhood of both standard and unknown are similar.

A sophisticated solution to the problem of background subtraction in scintillation spectroscopy involves the use of techniques (such as least-squares fitting) which determine the composition of complex mixtures by the shape of the entire γ-spectrum.

The same problems exist in principle with high resolution detectors, but the energy region covered by a photopeak is so much smaller that a linear interpolation of the underlying background is satisfactory. Most computer programmes developed for germanium detectors are in fact based on the simpler approach of linear background subtraction (Helmer et al., 1967; T. M. Wyckoff *private communication*; Guzzi et al., 1967). An exhaustive review of methods of computing experimental results in activation analysis has been given by Yule (1968). Often a photopeak from a Ge(Li) detector is described by a few points only. In this case the statistical accuracy in background subtraction is poor; it is well to have at least 8–10 points per peak. Sometimes an unduly low bias on the detector causes a tailing of the peak on the low energy side due to poor charge-carrier collection efficiency. Increase of the electric field improves the symmetry of the peaks. Planar drifted cylindrical crystals give more symmetrical peaks than coaxial detectors (Freeman and Jenkin, 1966) because of better definition of the electric field.

II. *Absolute γ-ray emission rates*

The absolute emission rate A_o of a γ-ray can be deduced from the experimental measurement of the photopeak counting rate A_p, through the relationship:

$$A_o = \frac{A_p}{E} \quad (4)$$

where E is the efficiency of the detector for that particular gamma ray. This efficiency can be found by calibrating the detector with sources of known gamma ray emission rates. In scintillation spectroscopy the efficiency can also be evaluated simply by means of the relation:

$$E = \epsilon_t P \quad (5)$$

Where ϵ_t is the absolute efficiency of the detector for a given energy, known from the literature and P is the peak-to-total ratio.

A correction factor to compensate for γ-ray scattering within the source and detector mounting should be applied when commercial detectors are used, or experimental peak-to-total ratios may be used, with certain limitations (Girardi et al., 1964).

With Ge(Li) detectors the use of relation (5) presents difficulties for the following reasons:

First, in scintillation spectrometry detectors have been standardized to a few preferred sizes. With Ge(Li) detectors there is still a great variety of dimensions and shapes, necessitating a search of the literature for the pertinent ϵ_t. Waino and Knoll (1966) have calculated the intrinsic efficiency for fully depleted cylindrical detectors having thicknesses of 1, 3·5, 8 and 12 mm and a surface area of 2·5 cm². Hotz et al. (1965) report ϵ_t for crystals of 2 cm² and 6 cm² surface area and depletion depths ranging from 1 to 10 mm with source-detector distances from 1 to 25 cm. Fry et al. (1966) and Decastro Fakia and Levesque (1967) report both photoelectric and double escape peak efficiencies, the former for cylindrical detectors of 18 mm dia. and depletion depths from 2 to 10 mm, the latter for crystals of depths 1 cm, 2·5 and 7·5 and the same diameters; sources were put at 10 cm distance from the crystal. Finally, Black and Grühle (1967) give efficiencies calculated for Ge(Li) detectors with sensitive volumes ranging from 0·3 cm³ to 8 cm³ and eight source-detector distances from 1 cm to 10 cm.

The calculated efficiencies have been compared by Fry *et al.* (1966) with experimental measurements reported by Ewan and Tavendale (1964), with good agreement.

Second, absorption and scattering from materials near the detector are more important for measurements on Ge(Li) detectors than for those on NaI(Tl) scintillators, due to the higher amounts of scattering materials near to the detector and the poorer geometry of measurement; correction factors should be calculated and applied.

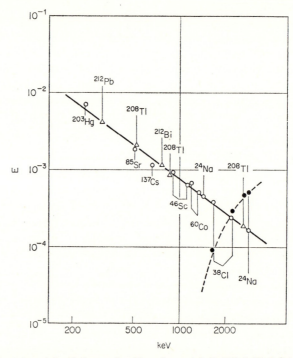

FIG. 3. Experimental efficiency of a 11 cm³ coaxial Ge(Li) drifted detector, against γ-ray energy: ○ values obtained with standardized sources. △ points obtained by means of a commercially available ²²⁸Th source. ● values for double escape peaks efficiency. (From Girardi *et al.*, 1967.)

Third, the determination of the photopeak area can sometimes present difficulties due to tailing effects, as already mentioned.

It seems therefore more practical, at present, to measure detector efficiency *versus* γ-energy by means of sources with known desintegration rates, or by comparison with a previously calibrated scintillator detector. Among the sources suitable for efficiency measurement the ²²⁸Th decay chain has the advantage of showing peaks in a large range of energy. Calibrated sources are commercially available and can be used for long periods of time due to the long half-life of the parent nuclide (1·9 years). The abundances of the various γ-rays of the source are given in Table II (Guzzi et al., 1966).

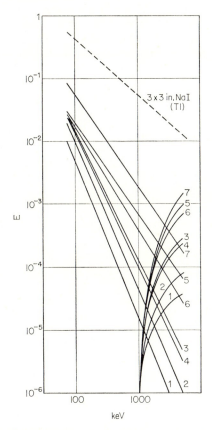

Fig. 4. Photopeak and double escape peak efficiency curves of seven Ge(Li) detectors, compared with a 3 × 3 in. NaI (Tl) crystal. (1) planar 0·625 cm³; (2) planar 0·8 cm³; (3) planar 1·2 cm³; (4) planar 1·8 cm³; (5) coaxial 10·5 cm³; (6) coaxial 18 cm³; (7) coaxial 47 cm³.

Figure 3 shows a typical calibration curve, obtained with ^{228}Th and other standardized sources.

In Fig. 4 the efficiency curves of seven detectors of increasing sizes, measured in our laboratory, are compared with the efficiency of a 3 × 3 in. scintillator. The increase of the detector size not only gives in general, a better efficiency (small variations are due to differing geometry of the counting position, which was particularly bad in detector 6) but also the dependence of efficiency *versus* γ-ray energy becomes less pronounced, due to the higher probability of total energy loss in the

TABLE II

Gamma Energies of ^{228}The Decay Chain and γ-Ray Abundances Adopted

Isotope	γ-Energy (keV)	γ-Ray Abundance (%)
^{212}Pb	239.36 ± 50.3	59.2
^{212}Pb	300.40 ± 0.32	4.0
^{228}Tl	510.40 ± 0.31	32.0
^{228}Tl	583.30 ± 0.29	88.0
^{212}Bi	727.7 ± 0.22	7.1
^{208}Tl	859.6 ± 0.29	14.0
^{208}Tl	1590.2 ± 0.70	double escape peak
^{208}Tl	2613.6 ± 0.30	100.0

sensitive volume of the detector by multiple Compton scattering with final photoelectric absorption.

Accurate measurements of the relative efficiency of Ge(Li) γ-ray detectors in the energy range 500–1500 keV, in a specific geometry, have been performed by Freeman and Jenkin (1966).

F. Applications to Activation Analysis

1. ANALYTICAL SENSITIVITY

Instrumental sensitivity of Ge(Li) detectors (that is, the possibility of detecting and measuring with a given precision a photopeak over a statistical background) has been studied and compared with that of NaI(Tl) crystals both by Pauly *et al.* (1966) and Lamb *et al.* (1966). The sensitivity depends on the efficiency of the detector, on the counting time and on the background under the peak. The absolute detection limit in γ-ray quantum per min. can be obtained by the following relation:

$$\nu = \frac{k^2}{2tE}\left(1 + \sqrt{1 + \frac{8t\nu_b}{k^2}}\right) \tag{6}$$

where the constant k depends on the confidence level required, t is the counting time in minutes, E is the efficiency of the detector and ν_b is the counting rate of the background in counts per minute.

Relation (6) has been applied to a 11 cm³ coaxially drifted Ge(Li) detector, both for pure cosmic background and when high energy γ-rays given by a ^{24}Na source were present in the spectrum (Pauly et al., 1966). In the first case it was shown that the 3×3 in. NaI(Tl) scintillator is about ten times more sensitive over the energy range 0·1–3 MeV.

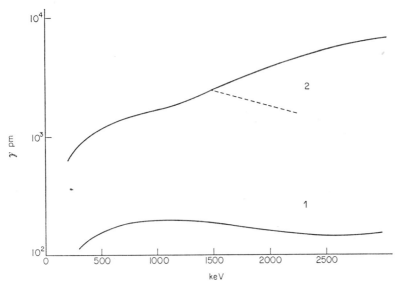

Fig. 5. The detection limits (as absolute γ-ray activity in γpm) over the natural background of a 3×3 in. NaI(Tl) crystal (curve 1), and of a 11 cubic cm coaxial Ge(Li) detector (curve 2) against γ-ray energy. The dotted line corresponds to the detection limit obtained if double escape peaks are measured. (From Girardi et al., 1967.)

In the presence of a ^{24}Na source the energy regions in which the scintillator is more sensitive (those with a flat Compton continuum) and those in which the Ge(Li) detector is more sensitive (in the neighbourhood of peaks) balance each other so that the amount of information on other activities that can be obtained with the two detectors is similar. Should more than one isotope be present in the source (as is often the case in activation analysis) the amount of information given by the scintillator would further diminish down to the point at which the γ-spectrum would be completely useless. In these cases the Ge(Li) detector would still give valuable information (see, for example, Figs 5 and 6).

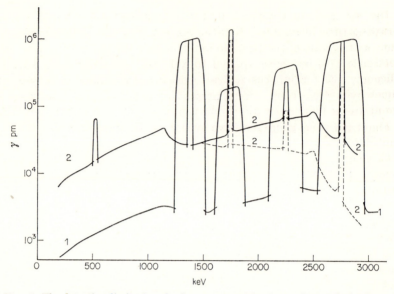

Fig. 6. The detection limits (as absolute γ-ray activity in γpm) over the background produced by a ^{24}Na source—disintegration rate 10^6 dpm—for a 3×3 in. NaI(Tl) detector (curve 1) and a coaxial 11 cubic cm Ge(Li) detector (curve 2) against γ-ray energy. The dotted curve corresponds to the detection limits obtained if double escape peaks are measured. (From Girardi et al., 1967.)

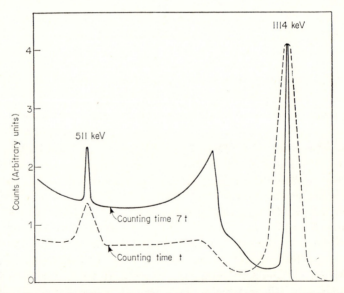

Fig. 7. γ-Spectra of a ^{65}Zn source: black line—coaxial 47 cubic cm Ge(Li) detector; dotted line—3×3 in. NaI(Tl) crystal.

Increasing the detector size increases the probability (for high energy γ-rays) of multiple Compton scattering with final photoelectric absorption within the detector. The peak-to-total ratio increases correspondingly, and the sensitivity becomes better. The advantage of higher efficiency and better peak-to-total ratio in large detectors is however somewhat balanced by inferior resolution, so that a large detector is not necessarily better for all problems than one of medium size.

Figure 7 shows the γ-spectra of a ^{65}Zn source measured on a 3×3 in. scintillator and on a 47 cm^3 semiconductor, normalized at the same peak height. The Compton continuum of the semiconductor is still appreciably higher than that of the scintillator, but the 511 keV peak is certainly as easily detected as with the scintillator.

Compton reduction, by surrounding the Ge(Li) detector with an annular plastic or inorganic scintillator to detect Compton-scattered photons and suppress them with a proper anti-coincidence circuit, has also been proposed to increase the detector sensitivity over Compton background (Cooper et al., 1967), is often done in scintillation spectrometry assemblies (Perkins and Robertson, 1965).

2. FIELDS OF APPLICATION

I. *High purity materials*

An application to the analysis of high purity materials was reported by Prussin et al. (1965), who determined concentrations of various elements in high purity aluminium. Short and long irradiations and measurements with different decay times allowed non-destructive determination of Cu, Mn, Ga, Na, Fe, Co, Cr, Hf and Sc. Girardi et al. (1965) applied Ge(Li) detectors to the determination of hafnium in zirconium oxide for nuclear use. The samples were irradiated for two hours in a thermal flux of 2×10^{13} n cm^{-2}s^{-1}. The crystal used in this work had an active volume of only 0·6 cm^3, but its high resolution made it possible to detect and measure the 0·133 and 0·482 MeV photopeaks of ^{181}Hf, obtaining a sensitivity of 10 ppm of Hf.

II. *Geological materials*

Applications to samples of geological interest have proved very valuable. Schroeder et al. (1966) determined silver in natural minerals. Intensities of the 660 keV γ-ray from 24 sec ^{110}Ag were measured after a 5 sec irradiation and a cooling time of 22 sec. The sensitivity reached was 15 ppm Ag. No chemistry was involved and the short irradiation time made it possible to reanalyse the same sample several times a day, to increase the analytical precision. Lamb et al. (1966) have applied

Ge(Li) spectrometry to the non-destructive analysis of a sulphide ore. Irradiation periods ranged from 2 to 72 hr in a thermal flux of 5×10^{12} n cm^{-2}s^{-1}. Ten elements were determined with the use of two Ge(Li) detectors of different active volumes (1·4 cm^3 and 6 cm^3) in the concentration range from $1·7 \times 10^5$ ppm (iron) to 2 ppm (gold). A comparative study on the determination of Mn was also performed by additional use of a 3×3 in. NaI(Tl) crystal. The analytical sensitivities obtained indicate that the larger Ge(Li) detector (6 cm^3 active volume) was capable of yielding sensitivities comparable to the NaI(Tl) crystal.

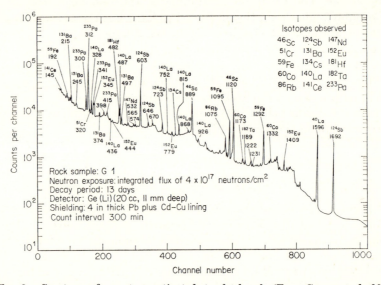

FIG. 8. γ-Spectrum of a neutron activated standard rock. (From Cooper et al., 1967.)

The non-destructive determination of lanthanides in rocks by Cobb (1967), is a successful application of the Ge(Li) detector to a problem which would have otherwise required long chemical separations. Eight different rock types were considered and seven rare earths (La, Ce, Sm, Eu, Dy, Yb, Lu) were determined as well as Sc, Mn and Th, which were the chief interfering elements in the measurements of the lanthanides. The Ge(Li) detector used was 1·6 cm^3 active volume. Relative abundances of the lanthanides found by Cobb fit well with data obtained by other methods, including chemical separations. Other applications to geological materials have been reported (Schroeder et al., 1966), including the analysis of the standard rock G 1, where 15 elements have been detected in one measurement (Cooper et al., 1967) Fig. 8, and the analysis of sea water (Cooper et al., 1967) Fig. 9.

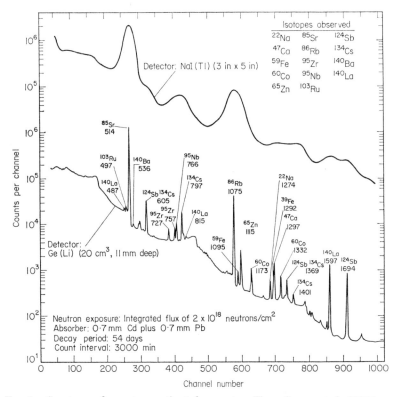

FIG. 9. γ-Spectrum of a neutron activated sea water. (From Cooper et al., 1967.)

III. Precious samples

Application to samples which must be preserved intact, either for their historical or archaeological interest, or for criminalistic problems are discussed by Schroeder et al. (1966). They analysed fifteenth-century printed documents in order to determine the constituents of moveable types used before 1540. Applications to forensic science of Ge(Li) detectors coupled to a large memory fast read-out analyser system have been reported by Bryan et al. (1967).

IV. Biological samples

Applications to biological matrices have been dealt with at the IAEA Symposium on Nuclear Activation Techniques in the Life Sciences held in Amsterdam in May 1967. Lyon (1967) presented a review of physical techniques of activation analysis in which the application of Ge(Li) detectors plays an important role. He obtained by means of a

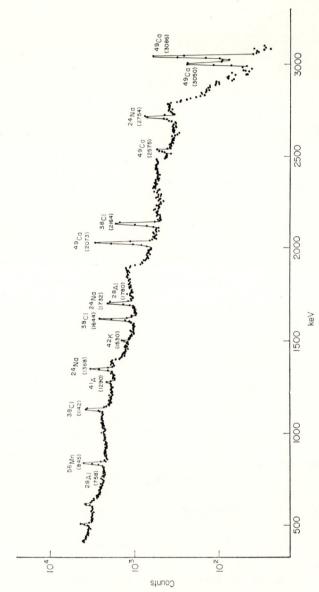

FIG. 10. γ-Ray spectrum of short-lived isotopes of a neutron activated biological sample: Irradiation time: 1 min.; Thermal neutron flux: 2×10^{13} n/cm^2 sec.; Decay period: 4 min; Count interval: 17 min; Detector: Ge(Li) drifted 10·5 cm^3. (From Girardi et al., 1967.)

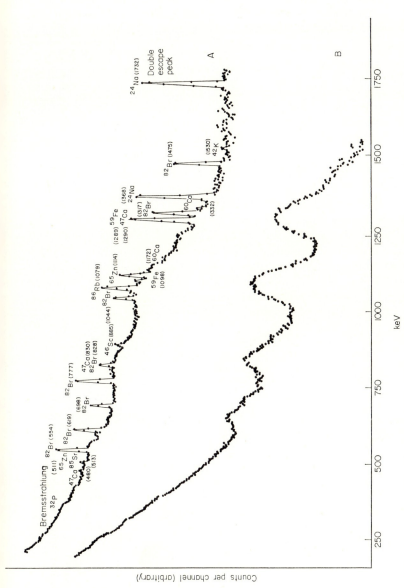

FIG. 11. γ-Ray spectra of long-lived isotopes of a neutron activated biological sample: Irradiation time: 8 dys; Thermal neutron flux: 5×10^{12} n/cm² sec; Decay period: 9 days; Count interval: 68 min; Detector: Ge(Li) drifted 10·5 cm³ (spectrum A); Spectrum B is the same sample counted for 1 min on a 3×3 (spectrum A); Spectrum B is the same sample counted for 1 min on a 3×3 in. NaI(Tl) crystal. (From Girardi et al., 1967.)

40 cm^3 crystal, γ-spectra of fall-out from atomic tests where a great variety of fission products was clearly identified. γ-Spectra of brain tissue taken with the same crystal were also examined. Girardi et al. (1967) used a biological reference material in a comparative study of two analytical techniques (activation analysis and X-ray fluorescent spectroscopy). Figures 10 and 11 show the γ-spectra of short and long lived activities. Routine measurements on biological samples from an ecological survey, including fish, algae, plankton, water, sediment, etc., have been also performed by Merlini et al. (1967) for the systematic determination of Zn, Fe, Co, Sr, Sc, Rb and Cs.

Tang and Tomlinson (1967) have used a Ge(Li) detector of 2 cm sensitive volume for the determination, of protein bound iodine. The peaks of ^{128}I at 338 keV and ^{87m}Sr at 442 keV were well resolved. Smith and Becker (1967) have determined Mn and Br in kale, beef-liver and white oak leaves, using a 4 cm^3 Ge(Li) detector. Perkins et al. (1967) with a semiconductor detector of 20 cm^3 having a peak-to-Compton ratio of 11-1 for the 1332 keV peak of ^{60}Co, have found ten different elements in human lung tissue in one measurement. Bromine concentrations in plants treated with pesticides have been determined by Wainerdi and Menon (1967) with a 2 cm^3 Ge(Li) crystal. Measurements of γ-spectra of blood were carried out by both Cooper et al. (1967) and Jervis and Wong (1967).

References

Bertolini, G. and Coche, A. (1968). "Semiconductor Detectors." North Holland, Amsterdam.
Black, J. L. and Grühle, W. (1967). *Nucl. Instrum. Meth.* **46**, 2, 213.
Black, W. W. and Heath, R. L. (1967). *Nucl. Phys.* **A90**, 650.
Bryan, D. B., Guinn, V. P. and Settle, D. M. (1967). *In* "Nuclear Activation Techniques in the Life Sciences". p. 61 I.A.E.A., Vienna.
Cappellani, F., Fumagalli, W., Henuset, M. and Restelli, G. (1967). *Nucl. Instrum. Meth.* **47**, 121.
Cobb, J. C. (1967). *Analyt. Chem.*, **39**, 1, 127.
Cooper, J. A., Wogman, N. A., Palmer, H. E. and Perkins, R. W. (1967). *BNWL* SA 1104.
Cooper, R. D., Linekin, D. M. and Brownell, G. L. (1967). *In* "Nuclear Activation Techniques in the Life Sciences", p. 65, I.A.E.A., Vienna.
Dams, R. and Adams, F. (1968). *Radiochim. Acta* **10**, 1.
Dearnaley, G. and Northrop, D. C. (1964). "Semiconductor Counters for Nuclear Radiations." Spon. London.
Decastro Faria, N. V. and Levesque, R. J. A. (1967). *Nucl. Instrum. Meth.* **46**, 325.
Ewan, F. T., Malm, H. L. and Fowler, I. L. (1966). Proc. IAEA Panel on Lithium-drifted Germanium Detectors, Vienna.

Ewan, G. T. and Tavendale, A. J. (1964). *Can. J. Phys.* **42**, 2286.
Freeman, J. M. and Jenkin, J. G. (1966). *Nucl. Instrum. Meth.* **43**, 269.
Fry, E. S., Palms, J. M. and Day, R. B. (1966). *LA* 3456.
Girardi, F., Guzzi, G. and Pauly, J. (1964). *Analyt. Chem.*, **36**, 1588.
Girardi, F., Guzzi, G. and Pauly, J. (1965). *Radiochim. Acta* **4**, 109.
Girardi, F., Guzzi, G. and Pauly, J. (1967). *Radiochim. Acta* **7**, 202.
Girardi, F., Pauly, J., Sabbioni, E. and Vos, G. (1967). *In* "Nuclear Activation Techniques in the Life Sciences", p. 229, I.A.E.A., Vienna.
Guzzi, G., Pauly, J., Girardi, F. and Dorpema, B. (1966). *EUR*, 3154 e.
Guzzi, G., Pauly, J., Girardi, F. and Dorpema, B. (1967). *EUR* 3469 e.
Heath, R. L., Black, W. W. and Cline, J. E. (1966). *Nucleonics* **24**, 5-52.
Helmer, R. G., Heath, R. L., Schmithtroth, L. A. and Cazier, G. A. (1967). *Nucl. Instrum. Meth.* **47**, 2305.
Hotz, H. P., Mathiensen, J. M. and Hurley, J. P. (1965). *Nucl. Instrum. Meth.* **37**, 93.
Jervis, R. E. and Wong, K. Y. (1967). *In* "Nuclear Activation Techniques in the Life Sciences", p. 137. I.A.E.A., Vienna.
Lamb, J. F., Prussin, S. J., Harris, J. A. and Hollander, J. M. (1966). *Analyt. Chem.* **38**, 7, 813.
Lyon, W. S. (1967). *In* "Nuclear Activation Techniques in the Life Sciences", p. 13, I.A.E.A., Vienna.
Mann, H. M., Bilger, H. R. and Shermann, I. S. (1966). *IEEE* NS 13-3, 253.
Merlini, M., Girardi, F. and Pozzi, G. (1967). *In* "Nuclear Activation Techniques in the Life Sciences", p. 615. I.A.E.A., Vienna.
Pauly, J., Guzzi, G., Girardi, F. and Borella, A. (1966). *Nucl. Instrum. Meth.* **42**, 15.
Perkins, R. W., Haller, W. A. and Torpe, J. D. (1967). *In* "Nuclear Activation Techniques in the Life Sciences", p. 557. I.A.E.A., Vienna.
Perkins, R. W. and Robertson, D. E. (1965). *In* "Modern Trends in Activation Analysis", p. 48. College Station, Texas.
Prussin, S. J., Harris, J. A. and Hollander, J. M. (1965). *Analyt. Chem.* **37**, 9, 1127.
Schroeder, G. L., Evans, R. D. and Ragaini, R. C. (1966). *Analyt. Chem.* **38**, 3, 432.
Schulze, W. (1962). "Neutronenaktivierung als Analytisches Hilfsmittel." F. Enke Verlag, Stuttgart.
Smith, G. W. and Becker, D. A. (1967). *In* "Nuclear Activation Techniques in the Life Sciences", p. 197. I.A.E.A., Vienna.
Tang, C. W. and Tomlinson, R. H. (1967). *In* "Nuclear Activation Techniques in the Life Sciences", p. 427. I.A.E.A., Vienna.
Tavendale, A. J. (1966). Proc. I.A.E.A. Panel on Lithium-drifted Germanium Detectors. *I.A.E.A.*, Vienna.
Taylor, J. M. (1963). "Semiconductor Particle Detectors." Butterworth, London.
Wainerdi, R. E. and Menon, M. P. (1967). *In* "Nuclear Activation Techniques in the Life Sciences", p. 33. I.A.E.A., Vienna.
Waino, K. M. and Knoll, G. F. (1966). *Nucl. Instrum. Meth.* **44**, 213.
White, D. H. and Groves, D. J. (1967). *Nucl. Phys.* **A91**, 453.
Yule, H. P. (1968). Int. Conf. Modern Trends in Activation Analysis, Gaithersburg, Maryland.

CLINICAL APPLICATION OF ACTIVATION ANALYSIS

D. COMAR

*Commissariat à l'Energie Atomique, Département de Biologie,
Service Hospitalier Frédéric Joliot, Orsay, France*

A. Introduction	163
B. Techniques in Activation Analysis of Biological Material	168
1. Samples	168
2. Irradiation	170
3. Chemical treatment	173
4. Measurement of radioactivity	176
C. Metabolic and Kinetic Studies	177
1. The relevance of trace element measurements	177
2. Kinetic studies	183
3. Stable tracers	188
D. Activation Analysis *in vivo*	192
1. General conditions	192
2. Whole body activation *in vivo*	194
3. Localized activation *in vivo*	198
E. Conclusion	202
References	202

A. Introduction

It is a remarkable observation that, of the 92 elements in the periodic table, fewer than 20 have been identified as normal constituents of biological tissue. Among these, oxygen, carbon, hydrogen and nitrogen account for 96% of the total mass of any organism, while calcium, phosphorus, sulphur, potassium, sodium, chlorine and magnesium make up about 3·6%. The remaining 0·4% is contributed by the 81 elements not already mentioned, classified under the general heading of trace elements. These elements may conveniently be divided into four categories, depending on the biochemical and physiological roles which have been found for them and on the concentrations at which they are present in biological material.

(a) Essential trace elements: iron (usually considered to behave both as a bulk and as a trace element), zinc, copper, manganese, iodine, cobalt, molybdenum and perhaps selenium.

(b) Trace elements always present in the body, for which an essential role has not yet been established; rubidium, bromine, barium, strontium, chromium, nickel, aluminium and arsenic.

(c) Elements thought to occur as contaminants but not found regularly in biological material: silver, lead, mercury.

(d) Elements which accumulate in the body during life: cadmium, tin, titanium, bismuth, gold.

The absence of information, conspicuous for the elements in the last three groups, is partly a consequence of the very low concentrations at which they occur in the body and of the corresponding difficulty of accurate estimation.

Table I shows the concentrations of the chemical elements in the soil, the whole body, the blood and the liver of the normal human subject. The liver has been chosen as an organ supporting intense metabolic activity, for comparison with the blood, which is the transport medium for trace elements.

For a large proportion of the elements in the periodic table, activation analysis is among the most sensitive analytical techniques now available. From this point of view it is the ideal method for studying the concentration of these elements in the normal tissues and for investigating changes in their metabolism associated with pathological conditions.

Among the criteria necessary to establish the essential role of a trace element, the demonstration of its invariable presence in an individual tissue of a particular species is of basic importance. The fulfilment of this condition can be claimed with confidence only when the analytical method used for measuring the element is of sufficient sensitivity. To take an example, mercury—which has been considered as a contaminating element—seems in the light of recent estimations by activation analysis (Kellershohn *et al.*, 1965) to be always present in blood. The concentration at which it occurs is, however, below the limit of sensitivity of conventional methods of microchemical analysis.

In the human red cell, the measured concentration of $0.0067\,\mu g/ml$ corresponds to 1,600 atoms of mercury per cell (Bowen, 1966). In order to detect a concentration of one atom per cell (and it is not possible to exclude the possibility that even this concentration may be of metabolic significance) the sensitivity of the estimation technique must be of the

TABLE I

Elemental composition of soil, human body, whole blood and the liver in the normal human subject

p.p.m	Soil (2 3 4 5 6 7 8 9)	Human Body (2 3 4 5 6 7 8 9)	Whole Blood (2 3 4 5 6 7 8 9)	Liver (2 3 4 5 6 7 8 9)
10^5	Si O	H C O	O	H C O
10^4	K Ca C Fe Al	P Ca N	N	N
10^3	Ti Na Mg	Na K S Mg	K S Cl P Fe	Cl Na P K Si Mg
	N	Cl	Na	Ca Fe
10^2	V Cr Zr Ba P S Mn	Rb Zn Fe	Mg Ca	
	Rb F Sr			
	Cl			Zn
10	Pb Cu Li Y	Rb Sr	Cu Rb Br Si Zn	Pb B Rb Cu
	B La Ni Ce	>As Cu Cd >La		Ni
	Ga Zn	>Pb Sn >Nb		
1	U Sb I Cs	>Sb Al	Se Pb As	Br Se Al Mn
	Ge Mo Br As Se Co	>Ni >Tc Ba Mn	Sn B F Al	Sn
	Th Be	B >Be >Co >Zr >Te		
10^{-1}	Tl Sc	>Li >Mo >Cr	Cr Ti Mn Ni I	Cd Hg
	Ag	>Au >Ag >Ru	Ag Li Sr	Sb Co
				Ag
10^{-2}	Hg Cd		W Sb Hg	Cr Mo Cs As
			Cs Mo Cd	
10^{-3}		>V >Bi	Co Ga	La
10^{-4}		>Cs U	Au	Au

In each box the classification of the elements represented on a logarithmic scale extending from left to right. For example the concentration of manganese in blood is 3×10^{-2} p.p.m. References: Bowen (1963), Brune and Samsahl (1966), Herring et al. (1960), Javillier et al. (1959), Parr and Taylor (1964), Samsahl and Brune (1965), Schroeder (1960), Spector (1956), Strain (1961), Tipton (1960), Widdowson and Dickerson (1964).

order of 10^{-13} to 10^{-15}g. Though activation analysis does not at present achieve this sensitivity, the development of reactors giving a very high flux of neutrons offers the prospect of approaching this limit in the near future.

It is of course pointless to attempt estimations at this level without the assurance of freedom from contamination by reagents used in the analytical procedure. Activation analysis allows this hazard to be avoided because reagents used during the estimation are not introduced until after the sample has been made radioactive. In these circumstances non-radioactive contaminants cannot interfere with the assay of induced radioactivity in the elements which are being determined. It is known that as analytical techniques improve and as reagent contamination is eliminated, the reported normal concentrations of trace element in blood tend to diminish. This process is quite well illustrated by recent changes in values reported for the normal concentration of manganese in plasma in the human subject (Table II). Without listing every published paper on the estimation of manganese in plasma, it is obvious that the levels found show a downward tendency.

TABLE II

Changes in reported concentrations of manganese in human blood plasma during the last 30 years

Year and Reference	Plasma Manganese μg per 100 cn³
Kehoe et al. (1940)	10·0
Bowen (1956)	2·4
Kanabrocki et al. (1964)	1·3
Bethard et al. (1964)	0·24
Cotzias et al. (1966)	0·059

Apart from the contamination peculiar to the analytical technique, the contamination produced *in vivo*, because of a diet rich in trace elements or through ingestion of medicaments, is often difficult to identify and to control. The existence of a regulatory mechanism (mitigating contamination of these kinds) may sometimes be inferred when systematic examination shows a uniformly low concentration of a particular trace element in blood samples from normal subjects differing in age and geographical origin. It seems that this regulatory mechanism is less severe for elements occurring at very low concentrations. Table III illustrates this point.

It is apparent that the body will react to a variation of more than $\pm 2\%$ in the level of sodium in the blood. On the other hand the physiological variability of trace elements appears to increase as their concentration diminishes. At present it appears that cobalt is the least abundant of the essential trace elements in blood. The concentration of cobalt in plasma is very low (0·28 μg/l). Only about 8% of this cobalt is present as a constituent of vitamin B_{12} (Parr and Taylor, 1964, Ekins and Sgherzi, 1965); the chemical state of the remainder of the plasma cobalt is still unknown.

TABLE III

Plasma concentration and relative physiological standard deviation of certain elements

Elements	μg/l	Relative standard deviation	Reference
Na	3250 000	2	Spector (1956)
Ca	99 000	10	Spector (1956)
Cu	1 190	11	Spector (1956)
Se	146	18	Tomlinson and Dickson (1965)
I	47	28·5	Comar and Kellershohn (1967)
Mn	0·59	31·2	Cotzias et al. (1966)
Co	0·28	57	Parr and Taylor (1964)

The foregoing considerations are relevant to the possibility of using activation analysis to study the distribution and metabolism of trace elements. For the study of metabolic processes it is necessary to determine simultaneously the concentration and the turnover rate of the element in the body. The scope of radioactive tracer techniques in this connection is limited in the human subjects by the necessity of avoiding radiation dosage beyond the limits generally considered acceptable for diagnostic tests. Furthermore, if the half-life of the appropriate isotope is too short, the information gained by tracer tests is incomplete and is limited to the more rapid metabolic phenomena. These difficulties may be avoided by using stable tracers, normally constituting a small proportion of the natural element. Mass spectrometry was the first technique used for studies of this kind, but its technical complexity has restricted widespread development. Activation analysis (which is essentially a method for the estimation of individual isotopes) is particularly useful for metabolic studies with an element which has a stable isotope of low natural abundance, capable of substantial enrichment. With such materials it is possible to measure the concentration in the body as a function of time after injection.

Finally, the third important aspect of activation analysis in relation to clinical science arises from the possibility of making non-destructive analysis. It is possible in principle (and to a useful extent in practice) to make direct measurements by γ-ray spectrometry of the concentrations of elements in a sample or even in the intact organism. This possibility has been exploited in recent work on the estimation of certain elements *in vivo*.

Before giving specific examples of the use of activation analysis in each of the three main areas which have been outlined, it will be useful to discuss a number of technical issues concerned with the preparation and handling of biological samples.

B. Techniques in Activation Analysis of Biological Material

1. SAMPLES

In procuring and preparing samples of biological material for the determination of trace elements by activation analysis, a number of precautions are necessary to avoid chemical contamination and other effects such as the volatilization and denaturation which may be produced in the course of the irradiation.

Biological fluids such as blood, urine and cerebro-spinal fluid are ideal materials for activation analysis. They are easily obtained and are automatically homogeneous in composition. Various sources of contamination and appropriate precautionary measures have been discussed by Bowen and Gibbons (1963). They recommend that blood samples should be handled in a way which avoids contact with glass in syringes or pipettes, because of the risk of contamination or loss by adsorption.

The use of anticoagulants should be also avoided. Bethard *et al.* (1964) showed that heparin contains $3 \cdot 56 \, \mu g/cm^3$ of manganese, $0 \cdot 65 \, \mu g/cm^3$ of copper and $28 \, \mu g/cm^3$ of zinc. A.C.D. (Acid-citrate-dextrose) may however be used for the separation of plasma as a preliminary to the estimation of these elements. Kellershohn *et al.* (1965) did not find measurable quantities of mercury in heparin.

Another kind of contamination may arise in the course of chemical treatment of the sample. Cotzias *et al.* (1966) have shown that the cellophane membrane used in dialysis of blood serum may release manganese in considerable amounts. They suggest that this effect may be responsible for the difference between their results and those obtained by other workers (Kanabrocki *et al.*, 1964) who did not observe the same phenomenon. Ion exchange resins are often used to separate the mineral forms of trace elements from the protein-bound forms. Comar *et al.* (1963) found that Dowex 1 and Dowex 2 anionic resins contained considerable amounts of iodine but that this element was not released during elution of the resin. The techniques of ultra-filtration and ultra-centrifugation used for the separation of proteins appear to be the least liable to contamination.

Though the blood is the means by which elements are transported in the body, it is in the tissues and more especially in the cells, where metabolic activity occurs, that their occurrence is especially significant.

Activation analysis is particularly appropriate for the estimation of trace elements in biopsy samples and equally in cellular fractions as has been proposed by Cotzias *et al.* (1964). By microdissection of thyroid tissue, Rivière (1968) was able to estimate iodine in single vesicules thyroidiennes with dimensions of the order of 100 μ.

The way in which samples are prepared for irradiation depends on the nature of the irradiating particles and on the available flux. If thermal neutrons are used at a flux not exceeding 4–$5 \cdot 10^{12}\, n\, \text{cm}^{-2}\, \text{sec}^{-1}$ and if the radiation time is not more than a few days it is possible to irradiate liquid samples sealed in polyethylene tubes. If blood serum or plasma is irradiated, coagulation of protein occurs after about an hour's irradiation. If the elements under investigation yield short-lived isotopes the time of irradiation can be kept short and the fact that the samples remain in the liquid state greatly facilitates any necessary chemical treatment after irradiation. If the neutron flux is greater than $10^{13}\, n\, \text{cm}^{-2}\, \text{sec}^{-1}$ and the irradiation time is long, it is advisable to use quartz containers and essential to dry the samples before irradiation, in order to avoid the release of gases associated with the decomposition of water by γ or neutron irradiation.

Contamination, and loss of the element under investigation, may be minimized by carrying out lyophilization of the sample in the irradiation tube. When this precaution is observed it is possible to estimate accurately a volatile element, such as mercury, in a blood sample.

When the neutron flux is still higher ($10^{14}n\, \text{cm}^{-2}\, \text{sec}^{-1}$) organic matter, even though dehydrated, may decompose and it is therefore necessary for the samples to be ashed before irradiation. The risks of evaporation and contamination may be reduced by ashing at a low temperature (below $100\,^\circ\text{C}$) in a current of atomic oxygen (Gleit *et al.*, 1962).

Standardization represents a significant problem in the estimation of trace elements. At the present time, very few standard biological materials (of animal origin) are available; those offered commercially do not provide accurately known concentrations of the less abundant trace elements of importance in chemical science. The only material of this kind now available consists of blood serum or urine, calibrated for iron, copper and iodine. Bowen (1967) recently prepared a vegetable standard from kale leaves (Brassica oleracea) dessicated and reduced to powder. Forty elements have been estimated in 29 different laboratories; the results show a wide variation of reported concentrations for several elements.

Standard aqueous solutions are the most often used, but they carry the possibility of substantial errors related to adsorption of trace elements on the surface of the container during irradiation. It has for example been shown by Comar and Kellershohn (1961) that half of the iodine in an aqueous solution of ammonium iodide (0·05 μg/cm^3) was adsorbed on the surface of a polyethylene tube after 20 mins irradiation in a flux of $5 \cdot 10^{12}$ n cm^{-2} sec^{-1}. This loss was avoided by the addition of sodium hydroxide (3g/l) to the solution. Mercury is also liable to adsorption on the wall of the container.

2. IRRADIATION

Irradiation in a nuclear reactor remains the most widely used method of activation, despite the continuing development of other sources of particles. The range of elements produced by thermal neutron bombardment of a sample of blood is illustrated in Fig. 1. The curves show the induced activity produced in an irradiation of eight days at a flux of 10^{13} n cm^{-2} sec^{-1}; the sample consisted of 1 cm^3 of normal blood. Because of the large concentration of sodium in biological materials and the large cross section of this element for thermal neutron capture, the greater part of the induced radioactivity is attributable to sodium-24. Figure 1 also show the changes, in induced activity of the elements in the sample during the course of decay.

The decomposition of the organic matter which constitutes the bulk of the sample has attracted the attention of biologists. This decomposition is attributable mainly to the substantial production of heat (several W/g) associated with the intense γ-ray flux which exists in the core of the nuclear reactor. Brune and Landstrom (1966) and Brune (1969) have described a method of refrigerating the sample during irradiation, allowing the maintenance of a temperature of $-40°$C during an irradiation of 13 hr in a flux of $2 \cdot 10^{12}$ n cm^{-2} sec^{-1}. This technique has the advantage of preventing coagulation of proteins and volatilization of certain trace elements as mercury and iodine. Gaseous constituents of the sample, such as dissolved argon, may also be estimated without loss.

Epithermal and fast neutrons, produced in pool type reactors, have been used for the selective activation of elements with a high cross-section for resonance capture. Borg et al. (1961) activated selectively the manganese in a sample of whole blood after removal of neutrons at energies below 300 eV by a filter of boron carbide (enriched in ^{10}B) surrounded by cadmium.

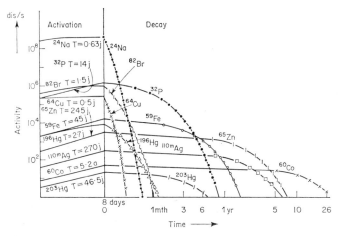

FIG. 1. Growth and decay of induced radioactivity in certain elements in a 1 cm^3 of normal blood.

For the estimation of trace elements by activation analysis, nuclear reactors are likely to remain the most acceptable radiation sources. A number of other possibilities are however beginning to be used. Equipment of modest size, which may be installed in a laboratory, is already available and is being used in chemistry and metallurgy for the estimation of light elements. Fast neutron generators, using the deuterium tritium reaction, can produce neutrons of energy 14 MeV at fluxes of the order of 10^9 n cm^{-2} sec^{-1}. Though these fluxes are not yet competitive with those delivered by reactors, a number of elements of biological interest (including nitrogen, oxygen, fluorine, phosphorus, sulphur, chromium, iron zinc and selenium) can be estimated at concentrations of the order of a few parts per million. Fast neutrons are interesting also because of their ready penetration of biological tissues.

This property was exploited by Anderson et al. (1964) to estimate total body sodium *in vivo* in the human subject.

Small cyclotrons for medical use are at present available commercially and their installation in hospitals offers many interesting possibilities. Simple to use and to maintain, machines of this kind will accelerate protons to 15 MeV, deuterons to 7·5 MeV and helium-3 ions to 20 MeV, with a beam current of a 100 μA. Cyclotrons of this sort are useful for the production of short-lived positron emitting isotopes, serviceable in a number of diagnostic applications. They are suitable also for activation analysis by fast neutrons, available at high fluxes from the reaction ^{9}Be $(d, n)^{10}$B, as well as by thermal neutrons (available at a flux of $5 \cdot 10^9$ n cm^{-2} sec^{-1} after moderation in paraffin wax) and by charged particles. Fleischer (1965) has calculated comparative sensitivities of estimation for all of the elements using activation by thermal neutrons, protons, deuterons, and helium-3 ions. Apart from the light elements, up to fluorine, for which activation by charged particles is much more sensitive than by thermal neutrons, there are a number of heavier elements of biological interest, such as calcium, sulphur and phosphorus, for which charged particle activation analysis is attractive. Fleckenstein et al. (1960), in studies on the turnover of A.T.P. in red cells, were able to estimate oxygen-18 by the reaction ^{18}O (p, n) ^{18}F using 4 MeV protons as the activating particles. More recently Vartapetyan et al. (1966) studied the incorporation of ^{18}O in vitamin A in the rat, using the reaction ^{18}O $(\alpha, n\gamma)$ ^{21}Ne, with α-particles of energy 4·6 MeV. Peisach and Pretorius (1966) have estimated ^{44}Ca and ^{48}Ca in blood by (p, n) reactions with 5 MeV protons, yielding the isotopes ^{44}Sc and ^{48}Sc. A major difficulty in the use of charged particles arises from their very poor penetrating power, necessitating the preparation of samples for irradiation in very thin slices. The analysis of superficial layers is of course sometimes important in the study of bone. Stubbins and Fremlin (1967) have examined the distribution of carbon, calcium and phosphorus in different layers of dental enamel by irradiation of samples with 2·2 MeV deuterons.

Some elements are not made radioactive by thermal neutrons or yield isotopes with such short half-lives that radioactive assay is not practicable without preliminary removal of sodium-24. In order to estimate these elements, without the complication associated with the abundant production of ^{24}Na, Cooper et al. (1967) performed activation analysis with 25 MeV bremsstrahlung produced by a linear accelerator. They were able to estimate carbon, oxygen, nitrogen, iodine, magnesium and barium in blood and in malignant tissues. Mulvey et al. (1965) estimated protein-bound iodine in blood after irradiation of samples with 22 MeV bremsstrahlung; the accelerator current was 250 μA and the irradiation time 90 min. Under these conditions a sensitivity of the order of 1 μg can be achieved.

3. CHEMICAL TREATMENT

The necessity for chemical treatment after irradiation arises from the great abundance of radioactive isotopes of chlorine, sodium and phosphorus after thermal neutron irradiation. In particular, some trace elements yield radioactive isotopes which, having half-lives shorter than those of ^{38}Cl and ^{24}Na, cannot be estimated by radioactive assay until these two interfering activities have been removed. If for example it is desired to decontaminate a sample of blood serum in order to estimate ^{128}I by beta counting without interference from other elements at a level exceeding 1% of the induced activity in ^{128}I, the decontamination factor for chlorine and sodium must be greater than 10^6. The methods commonly used for the chemical treatment of irradiated biological samples involve destruction of organic matter by mineral acids and separation by solvent extraction, precipitation or chromatography on ion exchange resins. The destruction of organic matter may take a considerable time and in some instances restricts the possibility of estimating short-lived activities. Though rapid methods of ashing based on fusion of the sample in a mixture of sodium peroxide and sodium hydroxide have been developed by Fukai (1959) it is better to avoid this stage altogether if possible.

In this respect some attention has been given to the possibilities offered by the Szilard-Chalmers effect for direct separation of trace elements from irradiated samples. After irradiation of liver samples, Brune (1966) extracted the following elements in concentrated hydrochloric acid, in the presence of carrier: ^{82}Br, ^{47}Ca, ^{60}Co, ^{51}Cr, ^{134}Cs, ^{64}Cu, ^{59}Fe, ^{24}Na, ^{86}Rb and ^{65}Zn.

Complete separation of ^{27}Mg from irradiated muscle has been achieved by the same procedure. Comar et al. (1967) have studied the radiolysis of organic compounds of iodine and of bromine during irradiation by thermal neutrons in a nuclear reactor. After irradiation, the radioactive bromine and iodine were completely extracted by a solvent or fixed on anionic exchange resins. Organic iodine and bromine appeared under these conditions to demonstrate Szilard-Chalmers effect. However, decomposition of the sample was partly attributable to the γ-radiation to which they were submitted. Samples of iodotyrosine were exposed to an intense flux of γ-radiation from a source of ^{60}Co and subjected to paper chromatography. The position of the spots indicated transformation of organic iodine into iodide. The application of these principles has led to the development of simple techniques for the estimation of magnesium in biopsy samples of a few mg (Brune and Sjoberg, 1965). Separation techniques applicable to particular elements in irradiated tissue samples have been described by several authors (Bowen and Gibbons, 1963; Lenihan and Thomson, 1965) as well as in the Symposium on Radiochemical Methods of Analysis (1965) and in two review papers (Girardi, 1969; Meinke, 1963).

The present trend is towards the development of completely automatic methods of clinical separation, allowing the estimation of an element in a large number of samples or the estimation of several

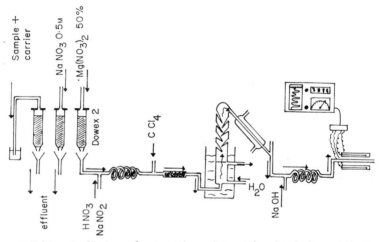

Fig. 2. Schematic diagram of automatic equipment for chemical manipulation in estimation of protein-bound iodine in blood by activation analysis.

elements in the same sample. Figure 2 shows the outline of a separation method for iodine in liquid biological samples (Comar and Le Poec, 1965). After irradiation for 30 min., the samples of serum or urine are moved automatically and transferred in succession to columns of anionic exchange resin. Halogens (including iodine), released from their organic attachment during irradiation are trapped by the resin while cations and proteins are removed by washing with water. Chlorine and bromine are then eluted by sodium nitrate (0·5M) and discarded. Iodine is finally eluted by magnesium nitrate (half saturation). After oxidation, the iodine in the eluate is extracted by carbon tetrachloride and distilled. The radioactivity of the distillate is estimated in a low background beta counter.

This apparatus, in which the operations are automatically programmed, allows one estimation to be made every five minutes. The sensitivity of measurements of iodine is routinely of the order of 0·001 µg. The same apparatus has been used to estimate bromine and plasma and urine with a sensitivity of 0·001 µg by means of the isotope ^{80m}Br. Apparatus of this kind is capable of very useful application in clinical science where activation analysis can compete with other methods only by performing rapid estimations with good sensitivity and precision. It seems that, despite the current development of automatic analytical equipment for use in biochemical laboratories activation analysis will retain a prominent place because of the accuracy and sensitivity that it provides. Other work on the automation of chemical separation has been done by Girardi et al. (1964), who have developed equipment allowing simultaneous estimation of several elements in a sample. The advantage of this equipment is in the flexibility of the design, allowing ready adaptation to a variety of problems.

Samsahl et al. (1963) have developed a series of chemical separation methods allowing the estimation of 30 elements in the same sample. In order to estimate simultaneously elements of long half-life, these workers divided the irradiation process into two steps—an irradiation of several days to produce elements of long half-life, followed by several days of decay to allow the disappearance of sodium-24 and a re-irradiation of short duration to re-establish the short-lived activities. Using a combination of distillation and chromatography on different ion exchange resins, Samsahl (1966) separated the elements into 16 or 18 groups (using wholly automatic methods for the last six groups) in a time of under an hour. Instead of absorbing the elements on a single ion exchange column and eluting them with reagents of different concentrations, Samsahl absorbed the elements in solution in a reagent of variable concentration passed in succession over a series of ion exchange columns. The variation of concentration was obtained by introducing the reagent at appropriate dilution between one ion exchange column and the next with the help of a peristaltic pump.

A significant feature of all of these automatic methods is the reproducibility with which they separate chemical elements, avoiding the necessity to measure the yield of the various operations. Somewhat similar principles underly the substoichiometric technique recently developed by Ruzicka and Stary (1964) which offers new possibilities for speeding up difficult chemical separations. Stary and Ruzicka (1964) have investigated theoretically the possibility of estimating about 25 elements; those of biological interest include As, Co, Cr, Cu, Fe, Mg, Zn. However, biologists are as yet little interested in these methods in relation to activation analysis. The volume of work carried out by analytical chemists since 1960 in trying to develop simple and reliable separation schemes or methods applicable to particular elements demonstrates the difficulty of this problem. However, the extensive application of activation analysis in relation to medical and biological problems requires that the advantages of the methods—sensitivity and absence of contamination—are not obliterated by the complexity of the means to attain them.

4. MEASUREMENT OF RADIOACTIVITY

The best analytical technique is one which requires the least manipulation of the sample. Instrumental activation analysis, in which the integrity of the sample is preserved throughout, has found only a few specialized applications in biology. The resolution of γ-ray spectroscopy of radioactive samples using a sodium iodide crystal is not comparable with that offered by spectrophotometry (arc, spark or flame) or by atomic absorption. The γ-ray spectrum of the sample of irradiated blood shows photo-electric peaks attributable to sodium and to chlorine; after the decay of these activities, isotopes of longer half-life (such as iron and zinc) may be distinguished. After the total decay of these elements it is possible in principle to estimate cobalt.

However, for the sample on which Fig. 1 is based, it would be necessary to wait for 5–10 years in order to make this final estimation.

The recently developed semi-conductor detectors have a poor sensitivity in comparison with sodium iodide crystals but have a very much better energy resolution which may be of decisive importance in the future application of instrumental activation analysis. The lack of sensitivity can to some extent be compensated by irradiation in a very high flux, of the order of $2-3 \times 10^{14}\, n\, \text{cm}^{-2}\, \text{sec}^{-1}$. With such fluxes, which are not unduly difficult to obtain at the present time, many elements may be estimated directly. Cooper et al. (1967) used a Ge (Li) detector to study the γ-ray spectrum of blood irradiated by thermal neutrons. A short time after the end of the irradiation, the photoelectric peaks of potassium, copper and manganese are visible despite the great abundance of sodium-24. After a period of decay of 10–12 days, the sodium and other elements of short half-life have disappeared, leaving photoelectric peaks attributable to tin, mercury, chromium, cadmium, gold, antimony, bromine, silver, rubidium, iron, zinc and cobalt. Linekin et al. (1969) used a coaxial Ge(Li) detector of volume 35 ml in their estimation of chromium, bromine, rubidium, iron and zinc in experimental tumours in mice. Tang and Tomlinson (1967) also reported the use of a germanium-lithium detector to estimate iodine in blood in the presence of other trace elements. Where it is important to obtain the highest sensitivity, γ-ray spectrometry with a sodium iodide detector (activated with thallium) remains the method of choice; by the use of a computer for decomposition of the complex γ-ray spectra, several elements may be estimated simultaneously. Without resorting to this degree of complexity, Dickson and Tomlinson (1967) were able to measure selenium in a variety of biological samples, using the isotope ^{77m}Se of half-life 17 sec. A simple dialysis preliminary to irradiation eliminated all of the unwanted elements and the spectrum obtained 30 secs after irradiation allowed the identification of ^{77m}Se and the estimation of selenium in the original sample with a sensitivity of the order of 0·005 μg. Some elements of biological interest cannot be estimated by conventional neutron activation analysis either because they yield, after (n, γ) reactions, isotopes of very short half-life or because they do not give radioactive isotopes at all in response to thermal neutron bombardment. Other elements, such as sulphur and phosphorus, though readily activated by thermal neutrons, give isotopes which are pure β-emitters and therefore cannot be measured by non destructive techniques. Comar et al. (1969) have exploited the capture γ-radiation (emitted from the compound nucleus immediately after incorporation of the bombarding neutron) to identify and in some instances to measure quantitatively a number of elements including hydrogen, boron, chlorine, sodium, potassium, nitrogen, sulphur and phosphorus. The lines in a typical capture γ-ray spectrum are closely spaced in energy and are best studied with a Ge(Li) detector. The resolution of γ-ray spectrometry may be greatly improved by the use of coincidence techniques. Keish et al. (1965) have described a coincidence equipment allowing the estimation of ^{129}I with extreme sensitivity, corresponding to a ratio ^{129}I/^{127}I of 3×10^{-13}.

C. Metabolic and Kinetic Studies

1. THE RELEVANCE OF TRACE ELEMENT MEASUREMENTS

Though activation analysis is certainly the best method for the estimation of a great many trace elements at the present time, there are still few examples of the successful application of the technique to diagnostic problems. Activation analysis often emerges unfavourably from comparison with other techniques for the estimation of essential trace elements such as iron, copper and zinc; for these elements, the correlation between concentration levels in biological materials and the corresponding pathological conditions is already well known. Leaving aside such elements as these, there are three areas of research in which current developments seem likely to lead to fruitful applications of activation analysis.

(a) Determination of normal concentrations of trace elements in biological tissues.

(b) Study of correlation between various chronic pathological states and trace element concentration in particular organs or tissues.

(c) Study of the distribution of particular elements as an aid to the understanding of metabolic processes.

The concept of normal concentration is difficult to define and should be considered in relation to the geographical origin of the samples and the age of the subjects from whom it is practicable to obtain samples. It is well known that certain elements accumulate in particular organs with increasing age—cadmium and mercury in the kidney, aluminium vanadium and titanium in the lungs, barium and strontium in the bones, tin in the intestines—but it is not known whether such accumulations are beneficial or detrimental. Conversely, it is sometimes necessary to enquire whether a deficiency of a trace element will be accompanied by a physiological disturbance. It has been shown (De Castro, 1949) that the ecological disturbances brought about in a region by the introduction of new agricultural techniques lead to important deficiencies in the indigenous population because of changes in diet. Dietary deficiency of animal protein is now considered one of the principal causes of malnutrition. The role of certain mineral elements such as calcium and iodine (as well as of vitamins) has been clearly demonstrated but no systematic work has yet been undertaken concerning the influence of trace elements on chronic disease. Activation analysis allows these problems to be approached in a realistic way. In this connection the analysis of hair or nail samples may provide a significant index of the trace element content of the diet. Strain (1961) has shown that the estimation of zinc in hair provides an indication of deficiency of this element and that the transfer of zinc from the blood to keratinous tissues, such as skin, hair and nails is influenced just as much by climatic conditions as by certain pathological changes such as those associated with burning. Study of the correlation between tumour growth and trace element concentration in the tumour and in neighbouring tissues has been in progress for a decade, since the initial work of Tietz *et al.* (1957). Work of this kind has been stimulated by consideration of the carcinogenic action of many elements. Apart from arsenic, which has long been notorious in this connection, beryllium, selenium, aluminium, chromium, iron, cobalt, nickel, zinc, silver, tin and mercury are known to be capable of tumour induction in man. In considering the mode of action of these elements Furst (1960) emphasized that carcinogenic substances have the property of readily forming compounds with metals or at least of being metabolized in such compounds. The observation that the concentration of many elements is significantly less in tumours than in neighbouring normal tissue is consistent with the suggestion of Warburg (1956) that the growth of tumours is related to the change from aerobic to anaerobic metabolism. In this connection it is known that the oxidases are metallo enzymes and that a change in the concentration of trace elements, resulting in inactivation of the enzymes, may bring about metabolic changes (Schroeder, 1960).

Most of the information so far available on trace element concentrations in tumours and neighbouring healthy tissues has been obtained by spectrography. Activation analysis is however well suited to work of this kind, since it offers the possibility of multi-element estimation in samples of small bulk, obtainable by biopsy. Samsahl and Brune (1965) have undertaken a systematic study involving the estimation of 18–30 elements in healthy and malignant tissue. In the two subjects whom they have so far studied, one with hypernephroma and one with adenocarcinoma of the large intestine, most of the elements estimated have been found at significantly lower concentrations in primary tumours and metastases than in the corresponding healthy tissues. These findings are consistent with results obtained by spectrography (Tietz et al., 1957) but cover a greater number of elements and provide encouragement for further studies in this direction.

The possibility that trace elements are significant in the etiology of coronary heart disease is supported by the work of Schroeder (1960) and Biörk et al. (1965) who found a negative correlation between the hardness of drinking water and the mortality from heart disease. Variations in the plasma levels of a number of elements such as copper, iron, manganese, zinc, nickel, molybdenum and boron have been reported by several workers (Hanson and Biörk, 1957, Kannabrocki et al., 1964, d'Alonzo et al., 1963, Wacker et al., 1956). Our knowledge of the pathology of cardiovascular disease has been augmented by the recent work of Wester (1965) on the estimation of 26 elements in heart tissue from normal subjects and from persons who had suffered coronary infarction. The comparison was made first between normal tissue and undamaged tissue from victims of coronary infarction and second between damaged and undamaged myocardial tissue from the latter group. In the first series, myocardial tissue from victims of coronary infarction showed a diminished concentration of copper and molybdenum and an enhanced concentration of arsenic and cerium. In the second series of comparisons, damaged tissue showed diminished levels of cobalt, cesium, potassium, molybdenum, phosphorus, rubidium and zinc and enhanced levels of bromin, calcium, cerium, lanthanum, sodium, antimony and samarium. Interpretation of these results is not obvious and may well require considerable further statistical study and comparison with other clinical tests.

A number of other studies have dealt with the concentrations of individual trace elements and the modifications occurring in various pathological conditions. Iodine is the element for which metabolic behaviour is best understood, though some aspects of the biosynthesis of thyroid hormones are still obscure. From the clinical point of view, the estimation of iodine in biological fluids (such as blood and urine) is of decisive significance, because of the information obtained in this way is in most cases sufficient to establish a diagnosis. However, the measurement of total plasma iodine is not an adequate index of thyroid function because of the different chemical form in which the element occurs—iodide, protein-bound hormonal iodine, protein bound non-hormonal iodine and free hormones. The second of these components which corresponds to about 77% of the total plasma iodine, gives the best indication of thyroid function. The concentration of this fraction and of total plasma iodine in normal subjects, estimated by activation analysis, is illustrated in Fig. 3.

The group of 219 subjects used in this study was strictly selected. Persons who, by questionnaire, clinical examination and by tracer studies with radioactive iodine, gave any evidence of abnormal iodine intake were not included. In these circumstances, total plasma iodine, having a distribution similar to that of protein-bound hormonal iodine, is a good index of thyroid function. On the other hand a diet rich in iodine will considerably increase the plasma iodine without a corresponding increase in the protein-bound hormonal iodine.

Current interest in the estimation of iodine in biological materials by activation analysis is reflected in a considerable number of recent publications (Wagner et al., 1961; Manney et al., 1961; Smith et al., 1964; Comar et al., 1964; Mulvey et al., 1965). Although chemical methods based on the catalytic action of iodine on the redox system arsenious anhydride-ceric sulphate are often used in hospital laboratories, the sensitivity is limited because of the reagent blank corrections which are inevitable. By these methods it is difficult to estimate with accuracy concentrations below $\cdot 01$ $\mu g/cm^3$. The sensitivity of activation analysis, being about 100 times better, allows a greater precision in the concentration range of physiological significance. Further, the absence of interference allows useful comparison to be made among estimations in different biological materials.

In France many thousands of estimations are made by activation analysis every year for clinical purposes and it seems that a laboratory devoted to this sort of work could be economically attractive.

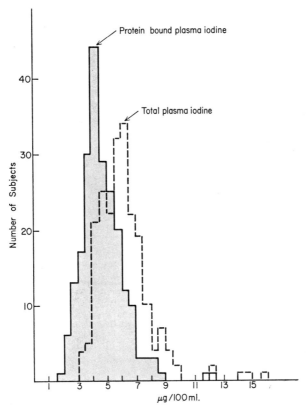

FIG. 3. Distribution of total plasma iodine and of protein-bound hormonal iodine in 219 normal subjects. Total plasma iodine = $6 \cdot 12 \pm 1 \cdot 9$ $\mu g/cm^3$ protein-bound hormonal iodine = $4 \cdot 74 - 1 \cdot 35$ $\mu g/cm^3$.

The main clinical problems in which estimation of iodine have been used are as follows:

(a) Control of the correction of iodine metabolism in hyperthyroid subjects treated by radioactive iodine (Comar et al., 1964).

(b) Discrimination between primary hypothyroidism and comparable conditions secondary to pituitary insufficiency, by estimation of protein bound iodine before and after administration of thyroid stimulating hormone (Rappaport et al., 1967).

(c) Study of iodine metabolism in children suffering from abnormalities of hormone production (Lissitzky et al., 1965; Rivière et al., 1967).

(d) Study of changes in concentration of protein bound non-hormonal iodine in patients suffering from thyroid cancer (Comar et al., 1963).

(e) Control of treatment of hypothyroidism by thyroid extract or thyroxine.

Activation analysis has recently been used for the estimation of selenium in biological material. Tomlinson and Dickson (1965), has shown that the concentration of this element in blood maintains a steady level and that in the plasma, selenium is mainly bound to α- and β-globulins. The part played by selenium and by vitamin E in the control of oxidation-reduction equilibrium in the tissues has recently been demonstrated in animals. The demonstration of physiological effects of deficiency of selenium in animals indicates the essential nature of this element (Rosenfeld and Beath, 1964). Some of the nutritional myopathies in man are attributable to deficiency of vitamin E (Horwitt, 1965) though association with modification of selenium metabolism has not yet been established. It was, however, shown by Duftschmid and Leibetseder (1967) that the chicken starved of vitamin E accumulates selenium in the muscle and the gonads. It will be seen that there are a number of research projects in which the elegant method for the non-destructive estimation of selenium, already described, may be particularly suitable.

Problems relating to the toxicology of mercury are of great current interest, particularly in view of possible hazards associated with the use of this element as compounds of pesticide for the treatment of cereals. The invariable presence of mercury in blood at a very low but relatively constant concentration (about $6\,\mu g/l$), as was shown by Kellershohn et al. (1965), suggested the existence of a regulatory mechanism. The distribution of mercury between the red blood cells and the plasma (similar to that of zinc), the chemical relationship of the two elements, the fact that both occur at high concentrations in the kidney (along with cadmium, which occurs in the same column of the periodic table) directs attention to the possibility of biological relationships between these metals. It is possible that the behaviour of these elements *in vivo* is linked with enzyme competition phenomena. Problems of the same kind arise in relation to molybdenum which has however been recognized as an essential trace element in plants and animals. Relationship between molybdenum (which occurs in certain enzymes, notably zanthine Oxidase) and other elements such as iron, tungsten and copper have been investigated in animals (de Renzo, 1962). Activation analysis, offering excellent sensitivity or the estimation of molybdenum (Livingstone and Smith, 1967) would allow similar studies to be pursued in the human subject.

Activation analysis extends the boundaries of trace element estimation, particularly when the available samples are too small for accurate estimation by traditional methods. Dubois et al. (1967) have described a method for the estimation of electrolytes (Na, K, Cl) and of phosphorus in tiny samples of muscle obtained by biopsy infants. This method has been applied to the study of acute disturbances of electrolyte in dehydration and milk fever as well as chronic conditions such as malnutrition and starvation. The results obtained appear to indicate the rapid involvement of muscle in water and electrolyte movement. In an earlier investigation, Bergström (1962) studied muscle electrolyte metabolism in normal human subjects and in patients suffering from renal insufficiency or from chronic diarrhoea, using biopsy samples of a few mg obtained with a new type of needle. More recently Woodruff et al. (1969) and Fite et al. (1969) have studied the possibility of diagnosing cystic fibrosis in the newborn by the estimation of sodium in nails. Non destructive activation analysis is convenient for serial measurements on large numbers of patients and, in association with other diagnostic tests, may give significant help in the early recognition of this very serious disease.

2. KINETIC STUDIES

Kinetic studies using the concept of body compartments require knowledge of the specific radioactivity of the element whose metabolism is being followed after injection of a tracer dose. Use of such methods in medicine depends on the availability of techniques sufficiently sensitive to measure the stable element and sufficiently specific to ensure that the estimations are comparable, in whatever tissue they are made. Activation analysis fulfil this criteria. Apart from kinetic studies made when the organism is in physiological equilibrium, investigations related to the disturbance of equilibrium by an external stimulus offers another possibility of studying regulatory mechanisms. From this viewpoint, the reaction of the organism to the administration of a significantly large dose of an element may advantageously be studied by activation analysis.

In thyroid physiology, activation analysis has in recent years made possible the study of movement of iodine between the different compartments in which it is distributed. Rivière *et al.* (1965), working with normal and hyperthyroid human subjects who were given a tracer dose of ^{125}I and daily doses of stable iodine, measured the specific radioactivity of iodine in urine, plasma, faeces as well as the uptake of the tracer in the thyroid over a period of a month. The four compartment model shown in Fig. 4 was deduced from the analysis of the results obtained in this investigation.

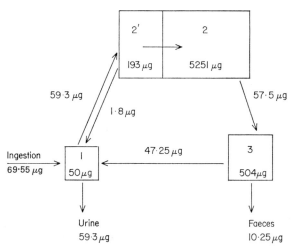

Fig. 4. Sizes of compartments and transfer rates for a 4 compartment model of thyroid function in a normal subject (Rivière *et al.*, 1965).
1—inorganic extrathyroidal iodine compartment;
2′—compartment containing inorganic iodine trapped in the thyroid and rapidly exchangeable organic thyroid iodine;
2—compartment containing slowly exchangeable organic thyroid iodine;
3—compartment containing extrathyroidal organic iodine.

This model comprises one compartment (1) of extra-thyroidal iodide, including dietary iodine and iodine produced by the breakdown of hormones, two thyroid compartments of which the first consists of rapidly exchangeable organic iodine and the second (2) of slowly exchangeable organic iodine, mainly incorporated in thyroid hormones, and a compartment (3) containing extra-thyroidal organic iodine representing the thyroid hormones in the circulating blood and other tissues.

The validity of this model is illustrated in Fig. 5 which shows calculated and observed values for specific activity of iodine. The model also demonstrates and indicates the extent of the secretion of iodide by the normal thyroid.

Rivière and his collaborators are now studying this release of iodine in patients with abnormal thyroid conditions as well as in cases where the thyroid has been blocked by anti-thyroid drugs or by ingestion of a substantial amount of iodine.

Comar et al. (1967) and Sklavenitis and Comar (1967) have studied bromine metabolism, particularly after administration of a dose of

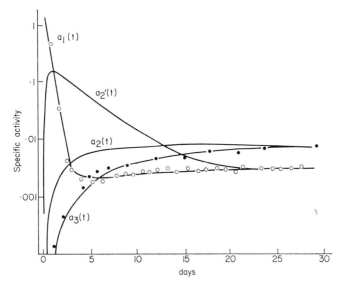

Fig. 5. Theoretical curves for specific radioactivity of iodine corresponding to the model in Fig. 4 (Rivière et al., 1965). The circles show the values found experimentally for the specific activity of iodine in the urine. The solid dots show the values found experimentally for the specific radioactivity of iodine in plasma hormones.

potassium bromide to normal and alcoholic subjects. Because of the short half life of bromine 82 (36 hr) dynamic processes studied with this isotope as a radioactive tracer can be followed for only a relatively short time and the incorporation of the element in protein cannot be demonstrated if its occurrence is slow. By electrophoresis of plasma proteins and estimation (by activation analysis) of bromine in the separated fractions, it has been shown that a small proportion of the order of 3%, of plasma bromine is in protein bound form. Figure 6 shows that protein bound bromine moves at the same rate as albumin.

After oral administration of a dose of potassium bromide (1g/50 kg body wt) the biological half life of bromine was found to be 11–12 days

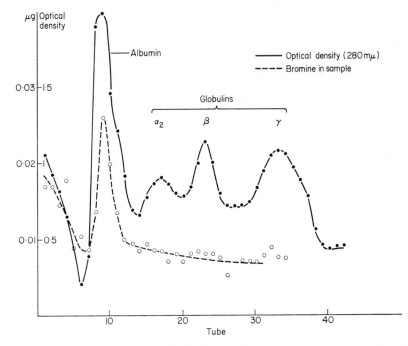

Fig. 6. Electrophoresis pattern of blood serum from a normal subject showing the distribution of natural bromine in various protein fractions (Comar *et al.*, 1967).

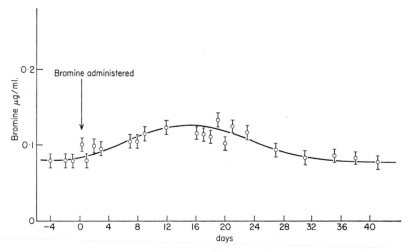

Fig. 7. Changes in the concentration of bromine in the plasma proteins of a normal subject before and after a dose of 1 g of Kbr (Sklavenitis and Comar, 1967).

in normal subjects and 9 days and in two subjects suffering from alcoholic cirrhosis. These results, or at any rate those obtained in normal subjects, are in agreement with the findings of Soremark (1960) who used bromine 82 in tracer amounts. The curve showing the disappearance of bromine from the plasma corresponds to a single exponential decline; this element is in fact eliminated almost entirely in the urine. Consequently a dose increasing the extracellular bromine space of the body by a factor of 10 does not seem to have any effect on the kinetics of distribution of this element. Evidence for the incorporation of bromine from the oral dose into plasma protein is seen in Fig. 7.

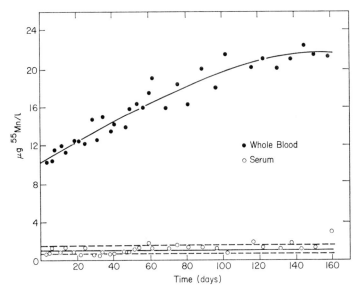

FIG. 8. Concentration in blood and in serum of manganese during continuous administration of the element (300 mg MnSO$_4$ per day) in a patient. The first day of dosage corresponds to zero on the time scale (Cotzias et al., 1966).

This uptake, which is small in extent and occurs slowly (but has been found to occur in several subjects) does not appear to be the result of passive adsorption, since the maximum concentration is produced 15 days after the administration of the oral dose.

Manganese is an element for which the extreme sensitivity of activation analysis has been a great advantage in metabolic studies. Cotzias et al. (1966) have studied the effect in blood manganese levels of a large oral dose of the element. Great amount must be ingested before even a modest increase in blood concentration level is observed. The time required to reach the new plateau corresponds approximately to the life of the red cells, as is shown in Fig. 8. This observation, considered in association with the slow decline in concentration after the end of oral administration of the element, suggests that manganese is incorporated in the haem of the red cells. Support is given to this suggestion by the observation that radioactive manganese is incorporated in the red cell porphyrins in animals.

3. STABLE TRACERS

A number of conditions are necessary for the useful exploitation of stable tracers in medicine and their estimation by activation analysis. It is necessary that the element whose metabolism is to be studied should possess a natural isotope of low abundance, that this isotope can be obtained at a high degree of enrichment and at an acceptable price and finally that the same isotope can be activated with a large cross section. For these reasons stable tracer techniques are limited mainly by the natural isotopic abundance of the element to be studied.

Oxygen-18, of natural isotopic abundance 0·02% may be obtained at a degree of enrichment of more than 90%. It should therefore be possible to label an oxygen-containing molecule with ^{18}O and to follow its behaviour in the body by activation analysis of the stable tracer. It is clear however that this tracer can be measured only for as long as it appreciably enhances the natural isotopic abundance of the element in which it is to be diluted. Oxygen-18 can be detected even when diluted 4000 times in natural oxygen. On labelling with ^{18}O a compound containing relatively little oxygen (such as cholesterol) it should be possible by using the method of isotopic equilibrium described by Chevallier (1966) to measure the rate of synthesis of this material in man. The sensitivity with which ^{18}O can be measured by the reaction ^{18}O (p, n) ^{18}F is enough to make an experiment of this sort theoretically possible.

Among the other elements present in organic matter, hydrogen may aptly be measured by stable tracer technique. Deuterium widely used as a stable tracer in association with mass spectrometry, may be detected quite readily by photonuclear activation (Odeblad, 1956) using γ-radiation of energy above the threshold level of 2·23 MeV. With a source of 15 curies of ^{24}Na, Wahl et al. (1966) expect to be able to measure 0·5 μg of deuterium. Table IV refers to a number of isotopes of small natural abundance which can be measured by activation analysis and which are of biological and clinical interest.

Several of these elements have already been studied in this way, Bethard et al. (1967) studied calcium metabolism in children after injection of calcium enriched to 31% in ^{46}Ca. Although the sensitivity of detection of this element is not so good as that offered by ^{48}Ca, the use of ^{46}Ca, which gives, by (n, γ) reaction, ^{47}Ca, with half life 4·5 days, simplifies the chemical manipulations. These authors were able to calculate for normal infants and for patients suffering from various

TABLE IV

Stable isotopes of clinical interest, capable of measurement by activation analysis

Isotope	Isotopic abundance (%)	cross section (barns)	Possible nuclear reaction	Half life of product
$^{2}_{1}$H	0·015	—	^{2}H$(\gamma, n)^{1}$H	—
$^{13}_{6}$C	1·11	0·2	^{13}C$(p, n)^{13}$N	10 min
$^{18}_{8}$O	0·204	0·5	^{18}O$(p, n)^{18}$F	1·87 hr
$^{26}_{12}$Mg	11·29	0·027	^{26}Mg$(n, \gamma)^{27}$Mg	9·45 min
$^{36}_{16}$S	0·017	0·14	^{36}S$(n, \gamma)^{37}$S	5 min
$^{41}_{19}$K	6·91	1·1	^{41}K$(n, \gamma)^{42}$K	12·52 hr
$^{46}_{20}$Ca	0·0033	0·25	^{46}Ca$(n, \gamma)^{47}$Ca	4·7 dy
$^{48}_{20}$Ca	0·1865	1·1	^{48}Ca$(n, \gamma)^{49}$Ca	8·8 min
	—		$(p, n)^{48}$Sc	44 hr
$^{50}_{24}$Cr	4·31	13·1	^{50}Cr$(n, \gamma)^{51}$Cr	27·8 dy
$^{58}_{26}$Fe	0·31	0·98	^{58}Fe$(n, \gamma)^{59}$Fe	45 dy
$^{64}_{28}$Ni	1·16	1·6	^{64}Ni$(n, \gamma)^{65}$Ni	2·56 hr
$^{70}_{30}$Zn	0·62	0·085	^{70}Zn$(n, \gamma)^{71m}$Zn	3 hr
			^{71}Zn	2·2 min
$^{74}_{34}$Se	0·87	26	^{74}Se$(n, \gamma)^{75}$Se	121 dy
$^{84}_{38}$Sr	0·56	1	^{84}Sr$(n, \gamma)^{85m}$Sr	70 min
			^{85}Sr	65 dy
$^{129}_{53}$I	—	20	^{129}I$(n, \gamma)^{130m}$I	9·2 min
			^{130}I	12·6 hr
$^{196}_{80}$Hg	0·146	3·100	^{196}Hg$(n, \gamma)^{197m}$Hg	24 hr
			^{197}Hg	65 hr
$^{204}_{82}$Pb	1·48	—	^{204}Pb$(n, 2n)^{203}$Pb	52 hr

abnormalities of bone formation, the total exchangeable calcium as well as the bone formation and resorption rates. McPherson (1965) used calcium 48 for studies of the same kind, with a chemical separation step to extract the calcium from tissue samples before irradiation.

Strontium enriched in ^{84}Sr may be used equally well for studies of this kind, as was proposed by Smith (1963).

In haematology, iron-58 and ^{50}Cr have been used to study the clearance of iron, to measure blood volume and to estimate the life of red cells, labelled with ^{50}Cr. The results obtained in these investigations are comparable with the findings of tests with radioactive tracers of the same elements (Lowman and Krivit, 1967).

^{129}I has been included in Table IV although this isotope may be used as a stable or as a radioactive tracer. It therefore offers a number of characteristics which make it an ideal tracer for clinical use (Edwards, 1962). By virtue of its chemical properties it may be used for *in vitro* labelling of any molecule including a tyrosine group. Labelling *in vivo* allows the study of thyroid metabolism. ^{129}I emits γ-radiation (38 keV, almost totally converted) which may be detected by external counting over the thyroid in a human subject after administration of a small dose.

Because of its very long half-life (and correspondingly low specific activity) it might be thought that the mass of ^{129}I necessary to make an adequate tracer dose would be so large as to interfere with the normal processes of iodine metabolism in the body. This difficulty need not arise because ^{129}I is readily activated by thermal neutrons to give two radioactive nuclides ^{130m}I (half-life 9·2 mins) and ^{130}I (half-life 12·6 hr) (Kellershohn *et al.*, 1967). The sensitivity of detection of ^{129}I by activation analysis and assay of ^{130}I is of the order of $5 \cdot 10^{-12}$g, using a thermal neutron flux of 10^{13} n cm^{-2} sec^{-1} and an irradiation time of 12 hr.

Irradiation of 10 min., followed by rapid chemical separation with ion exchange resin, allows simultaneous measurement of ^{127}I by the photo-electric peak of ^{128}I at 450 keV and ^{129}I by the photoelectric peak at 540 keV of ^{130m}I. A technique of this kind is particularly significantly important in kinetic studies where it is necessary to measure the specific activity of iodine in biological material.

Because of the unusual range of properties just mentioned, ^{129}I may be used simultaneously as stable and a radioactive indicator, allowing the prolonged study of thyroid metabolism. These considerations are illustrated in Fig. 9 which shows changes in the concentration of ^{129}I in the thyroid, the blood and the urine of a euthyroid subject after administration of a single dose of 325 μg of this element. The thyroid uptake was measured by external counting of ^{129}I and the concentration in blood and urine was estimated by activation analysis.

These measurements were continued for 260 days and could well have been pursued for several months longer. Rivière *et al.* (1969) using the occupancy principle described by Orr and Gillespie (1968) were able to estimate the ^{127}I content of the thyroid gland and the daily dietary intake of stable iodine in a normal human subject. These quantities were calculated by the application of the occupancy principle to data already available in the form of activity-time curves for the thyroid and the plasma after administration of a tracer dose of ^{129}I.

Because of the very small radiation dose given by ^{129}I to the thyroid gland (less than 1 rad in the experiment just described) it is possible also to use the isotopic equilibrium technique. This method, effectively used by Simon (1964) in studying thyroid metabolism in animals, has not until now been undertaken in man because of the great amount of

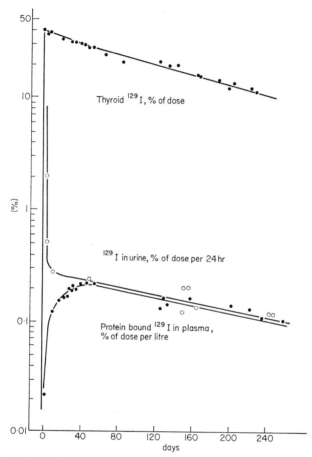

FIG. 9. Thyroid uptake, urinary excretion and protein plasma concentration of iodine-129 followed for 260 days in a normal subject given a single dose of 325 micrograms of iodine-129 (Comar and Kellershohn, 1967).

radioactive iodine which would be necessary. In contrast, a human subject receiving a daily dose of 10 μg (1·7 nCi) of ^{129}I for a year will incur a thyroid dose of 8 rads which is considerably less than that delivered by a single tracer dose of ^{131}I as normally used for diagnostic purposes. Calculation shows that a dose of more than 1000 rads would be incurred by the thyroid in any attempt to achieve isotopic equilibrium in man with ^{125}I (Rivière, 1968).

The isotopic equilibrium technique with ^{129}I was used in the rat by Ouelette et al. (1966) to study the form in which iodine occurs in the blood. Their identification of the amino acids which are iodinated to a substantial extent agreed with the results of Dimitriadou et al. (1964) but not with the findings of Simon (1964).

As well as the isotopes mentioned in Table IV, a number of other elements may be used as stable tracers. Some nuclides useful in this connection are not strictly speaking isotopes of the elements whose metabolism is being investigated but will serve the same purpose because of closely similar chemical and physiological properties. Bromine-82 is for example widely used to estimate the chlorine space in the body, or in other words the extra-cellular volume. Sklavenitis and Comar (1967) made isotopic dilution studies by activation analysis of blood and urine after administration of a single tracer dose of stable bromine (3 mg/kg body weight in the form of potassium bromide, enough to increase the plasma bromine level in a normal subject by a factor of 4). The validity of this technique for the estimation of extracellular volume was confirmed by parallel studies with ^{82}Br.

Bromine has been used as a stable tracer by Fireman et al. (1966) to label human proteins and to measure their biological half-life after injection into mice.

In the light of these investigations it is reasonable to expect that rubidium, which occurs to a very small extent in biological fluids, may be used for potassium.

D. Activation Analysis *In Vivo*

1. GENERAL CONSIDERATIONS

The potentiality of activation analysis for non-destructive tests are currently exploited in the estimation of amounts of elements *in vivo*. This possibility is of considerable interest physiologically because there is at present almost no way of determining the elemental composition of a living organism. Methods most frequently used are based on the dilution of radioactive tracer, allowing estimation of the exchangeable mass of an element or of a molecular species. This dilution in the body is seldom complete notably for electrolytes and for bone calcium where very slow turnover processes are involved. At other sites in the body (for example in the thyroid) the dilution of the tracer takes place in several stages corresponding to participation in a complex metabolic cycle. In these circumstances the attainment of the transitory isotopic equilibrium which allows the estimation of intrathyroidal iodine requires such a long time that the method can seldom be used effectively.

With these considerations in mind, it is apparent that the comparison of two methods, one allowing the estimation of total mass of an element and the other of exchangeable mass, can be very fruitful, particularly in identifying and measuring some important diagnostic parameters.

Studies of the radioactivity induced in human subjects after nuclear accidents suggested the possibility of producing measurable activity *in vivo* after exposure to a very modest neutron flux. Maletskos *et al.* (1961, 1963) proposed that bone calcium might be measured after irradiation of a finger by reactor neutrons.

In the light of theoretical and practical considerations a number of conditions must be satisfied if the technique of activation analysis *in vivo* is to be successful in human subjects.

It is important that the radiation dose delivered to the patient should be small and should be at the most, of the same order as that generally acceptable in a diagnostic test with radioactive tracers. For this reason the maximum flux used for *in vivo* irradiation is limited by the dose which the irradiating particles will deliver to the body. It is therefore feasible to detect only elements of adequate abundance in the body or those which are concentrated in a particular organ.

The more important elements which can be detected and measured by activation analysis in the human subject are shown in Table V. The feasibility and sensitivity of the technique for a particular element is influenced by the energy of the neutrons available for irradiation.

In experiments on activation *in vivo* made with a collimated beam of neutrons from the EL 3 reactor at Saclay, Sklavenitis *et al.* (1969) studied the distribution of dose on the surface and in the interior of tissue phantoms. A cylindrical phantom of tissue equivalent material was used to simulate the leg or neck of a human subject. A thermal neutron flux of $1 \cdot 25 \, . \, 10^8 \, n \, cm^{-2} \, sec^{-1}$ in the form of a collimated beam of diameter 3 cm was used for irradiation. The absorbed dose on the surface of the phantom was found to comprise the following components:

(a) Capture γ-rays coming mainly from the γ reaction on hydrogen-180 mrem/min.

(b) Recoil protons from the reaction $^{14}N(n, p)^{14}C$-1190 mrem/min.

(c) Direct γ-radiation and fast neutrons-94 mrem/min.

Inside the tissue phantom, the absorbed dose decreased rapidly because of the heavy absorption of thermal neutrons. At a depth of 5 cm, the dose due to capture γ-rays was approximately 10% of that recorded at the surface and the dose due to recoil protons was only 1% of the surface value. Absorbed doses measured by a number of authors during activation analysis *in vivo* are shown in Table VI, taken from the unpublished proceedings of a conference on *in vivo* activation analysis held at the Service Hospitalier Frédéric Joliot, Commissariat à l'Energie Atomique, Orsay, France in July 1968.

TABLE V

Estimation of certain elements by neutron-induced reactions *in vivo*

Stable element	% by weight in human body	Induced nuclide	Neutron reaction	γ-Ray to be measured
H	10	^{2}H	n, γ (thermal neutrons)	capture γ (2·2 MeV)
N	3	^{13}N	$n, 2n$ (14 MeV)	delayed γ (0·51 MeV)
Ca	1·5	^{49}Ca	n, γ (thermal neutrons)	delayed γ (3·10 MeV)
			n, γ (thermal neutrons)	capture γ (many)
P	1	^{28}Al	n, α (14 MeV)	delayed γ (1·78 MeV)
		^{32}P	n, γ (thermal neutrons)	capture γ (0·08 MeV)
Na	0·15	^{24}Na	n, γ (thermal neutrons)	delayed γ (2·75 MeV)
Cl	0·15	^{38}Cl	n, γ (thermal neutrons)	delayed γ (1·6-2·2 MeV)
		^{38}Cl	n, γ (thermal neutrons)	capture γ (many)
		^{37}S	n, p (14 MeV)	delayed γ (3·10 MeV)
Mg	0·05	^{26}Mg	n, γ (thermal neutrons)	delayed γ (0·84 MeV)
		^{24}Na	n, p (14 MeV)	delayed γ (2·75 MeV)
I	0·03 (thyroid)	^{127}I	n, γ (thermal neutrons)	delayed γ (0·450 MeV)

As might be expected, the radiation doses necessary to produce satisfactorily measurable levels of induced activity are greater for individual organs or limited regions of the body than for the entire body. It will however be seen from the last column of Table VI that the radiation doses involved in total or partial body irradiation for activation analysis are of the same order of magnitude as those generally accepted in diagnostic radiology or in tracer studies with internally administered radioactive substances emitting β-rays or γ-rays.

2. WHOLE BODY ACTIVATION IN VIVO

The first systematic analysis of problems involved in study of human subjects by activation analysis *in vivo*, and the first practical realization of the technique, were made by Anderson *et al.* (1964).

TABLE VI

Main features and dosimetry of recent *in vivo* activation analysis projects

Source of neutrons	Energy MeV	flux $n\ cm^{-2}\ s^{-1}$	irradiation	time of irradiation min.	absorbed dose rem	Authors
Cockcroft-Walton generator $^3H(d,n)^4He$	14·4	10^6	whole body man	4–7	1	Anderson et al. (1964)
Cyclotron $^7Li(p,n)^7Be$	3·5	10^6	whole body man	5	0·9–1·7	Chamberlain et al. (1968)
Cyclotron $^9Be(d,n)^{10}B$	2		whole body cadaver		0·38	Palmer et al. (1968)
Reactor	thermal	$1·25 \times 10^8$	neck sheep	10	14 (surface)	Lenihan et al. (1967) Sklavenitis et al. (1969)
Reactor	thermal	2×10^7	neck man	5	4–15	Boddy et al. (1968)
Reactor	cold	$2·7 \times 10^7$	leg man	5–30	2–12 (surface)	Comar et al. (1968)

A major difficulty arises from the problem of homogeneity of irradiation. If meaningful results are to be obtained in any activation analysis procedure it is desirable that the probability of the relevant nuclear events should be the same throughout the irradiated object or, at any rate, that the behaviour of the irradiated object in this regard should be closely similar to that of a suitable standard of known chemical composition.

Neutrons have limited penetrating power in tissue. Smith (1962) showed that it is impossible in practice to obtain homogeneous irradiation of the human body in a single exposure to neutrons, whatever their energy. Battye et al. (1967) showed that it is however possible to obtain reasonable homogeneity of irradiation of the entire body by using 14 MeV neutrons partially moderated by a blanket of polyethylene 3 cm thick wrapped around the experimental subject. The irradiation is conducted in two parts, the neutrons being directed first at the front and then at the back of the body. In these circumstances the flux of thermal neutrons, resulting from slowing down of the 14 MeV neutrons in the polyethylene blanket or in the tissues of the body, varied by a maximum of 3·5% for a body thickness of 25 cm.

On the other hand the fast neutron flux (measured by the activation of magnesium detectors) was found to be very inhomogeneous. For this reason it is difficult to achieve accurate measurements using fast neutron reaction such as $^{14}N(n, 2n)^{13}N$ and $^{31}P(n, \alpha)$, ^{28}Al.

Palmer et al. (1968), Chamberlain et al. (1968a) and Newton et al. (1969) obtained similar results with neutrons of mean energy between 2 and 3·5 MeV, produced by a cyclotron. In order to obtain homogeneity of irradiation, the experimental subject should be placed at as great a distance as possible from the cyclotron; in practice, the necessity of obtaining an adequate neutron flux has usually limited the separation of source and subject to a distance of about 2 m. In these circumstances a thermal neutron flux of the order of 10^6 n cm^{-2} sec^{-1} can be obtained at the surface of the body, allowing a few nCi of ^{49}Ca, ^{24}Na and ^{38}Cl to be produced during an irradiation of a few minutes.

The γ-ray spectrum obtained after irradiation *in vivo* is rather complex because of the simultaneous production of many activities. Figure 10 shows the γ-ray spectrum obtained (by measurement in a whole body counter) after neutron irradiation of a normal human subject.

Apart from ^{137}Cs (from fall-out) and naturally occurring ^{40}K, ^{38}Cl, ^{24}Na and possibly ^{13}N are evident in the spectrum. Estimation of total body sodium and chlorine may therefore be attempted. As the spectrum shown in Fig. 10 was recorded 30 min after the end of irradiation the photopeak associated with ^{49}Ca (3·09 MeV) is not visible. This activity was however measured with reasonable accuracy by Palmer et al. (1968) using a cadaver and by Chamberlain et al. (1968a) after irradiation of a living subject.

The calculation of elemental composition by the measurement of photopeak areas implies that the induced activities in the irradiated subject and in an appropriate standard can be measured with the same efficiency. This condition will be satisfied if the induced activity for a

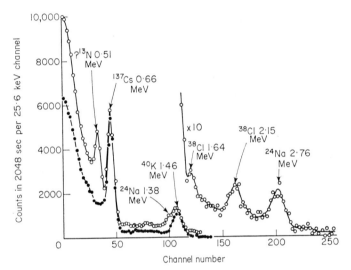

FIG. 10. γ-Ray spectrum from a normal subject (crosses) and approximately 30 min after (circles) irradiation by 14 MeV neutrons, partially moderated, to an absorbed dose of 0·1 rad (Anderson et al., 1964). (Reproduced by permission of the Lancet.)

particular element is distributed in the same way through the standard and through the irradiated subject. A reasonable approximation to this requirement can be expected for elements such as chlorine and sodium, of which the major part is distributed through the extracellular fluids of the body–though it must be remembered that about 30% of the sodium in the body is contained in bone, at a concentration approximately twice as high as that prevailing, on average, in other tissues.

A different situation is found for calcium, which is largely concentrated in the skeleton and for which it is difficult to construct a standard having the same geometry as the human body. It is for this reason that Chamberlain et al. (1968a) expressed their results for the estimation of body calcium in terms of induced activity rather than of mass.

Since however the main interest in the estimation of this element is associated with the management of diseases such as osteoporosis, repeated examinations on the same subject at intervals of time may be expected to give information of value to the physician.

The results so far obtained have involved normal subjects. It may however be reasonably claimed that the method has already been developed to the point of clinical usefulness. Estimation of total body sodium on three subjects has for example confirmed the existence of a non-exchangeable compartment in bone, already in evidence from isotopic dilution measurements (Chamberlain et al., 1968b). It seems, though, that the more immediate uses of whole body activation analysis will be in relation to the estimation of calcium and the variations of total body calcium in bone disease. Present indications are that a variation in total body calcium of the order of 2% may be reliably detected.

Though most of the studies so far reported have been concerned with the metabolism of mineral elements, Nagai (1968) has estimated total body nitrogen in mice as an aid to the investigation of protein metabolism. He used 14 MeV neutrons in the reaction ^{14}N$(n, 2n)^{13}$N. The γ-ray spectra of the irradiated animals showed a substantial photopeak corresponding to the annihilation radiation of ^{13}N at 0·510 MeV. The 1·78 MeV peak of ^{28}Al produced in the reaction ^{32}P (n, α) ^{28}Al was also evident. It should therefore be possible to estimate total body phosphorus by fast neutron activation.

Hydrogen, one of the major constituents of living matter, has been measured in the human subject by activation analysis *in vivo*. Rundo and Bunce (1966) observed that the γ-ray spectra obtained from un-irradiated normal subjects studied in a sensitive whole body counter usually included the 2·23 MeV γ-radiation produced in the reaction ^{1}H (n, γ) ^{2}H, with neutrons released by the interaction of cosmic radiation with the lead shield of the counting room. Further experiments in which a radium-beryllium neutron source was placed close to the whole body counter allowed more accurate estimation of total body hydrogen. An average value of 9·47% of body weight was found in 27 male subjects and a value of 8·65% in 10 female subjects.

3. LOCALIZED ACTIVATION IN VIVO

Activation analysis by localized irradiation *in vivo* is subject to the same difficulties as those encountered in whole body studies, particularly in regard to the homogeneity of irradiation. The technique is however of considerable interest when it is desired to measure the concentration of an element in a particular organ or to study the kinetic behaviour of a natural element by labelling with a stable tracer *in vivo*.

The thyroid is the most attractive organ for work of this kind because it includes, in a volume of about 30 ml in the normal subject, almost the whole of the iodine in the body. Furthermore it is situated conveniently close to the skin, allowing the use of neutrons of very low energy, thereby limiting the volume of tissue irradiated at a significant dose level. In experiments involving *in vivo* irradiation of sheep, Lenihan et al. (1967, 1968) used an internal standard to allow for the heavy absorption of neutrons in the tissues and to correct for the resulting inhomogeneity of irradiation.

This method allows the estimation of the ratio of probability of activation and of detection *in vivo* and in the standard. ^{129}I was used as an internal standard. After injection of a known quantity of this nuclide it is possible to estimate the mass trapped in the thyroid, this being the difference between the quantity ingested and the quantity eliminated in the urine during the two succeeding days. External counting of the radioactive emission from ^{129}I allows the exact position of the thyroid gland to be determined.

Under irradiation by thermal neutrons, ^{129}I is partly transformed into ^{130m}I and ^{130}I at the same time as ^{127}I is converted to ^{128}I. The ratio of the activities of ^{130m}I (or ^{130}I) and ^{128}I is measured in the thyroid *in vivo* and in a standard containing known quantities of ^{127}I and ^{129}I. It is then possible to calculate accurately the mass of ^{127}I in the thyroid without regard for the differences of neutron flux gradient in the tissues and in the standard.

The irradiations were made with a collimated beam (3 cm in diameter) of thermal neutrons with minimum contamination by fast neutrons and γ-radiation. After an irradiation period of 10 min, the induced radioactivity was measured with a sodium iodide crystal 20 cm in diameter and 10 cm thick. This crystal was surrounded by a lead collimator, 4 cm thick, in the form of a cylinder surmounted by a cone; this arrangement ensured that the response of the crystal was confined to γ-rays coming from the irradiated region.

Figure 11 shows the γ-ray spectra obtained under these conditions after irradiation of the neck of a sheep *in vivo* by thermal neutron flux of $3 \cdot 10^8$ n cm^{-2} sec^{-1}. Apart from the peak at 450 keV due to ^{129}I, there is a noticeable peak of 540 keV attributable to ^{130m}I produced by the ^{129}I used as an internal standard. The lower γ-ray spectrum shows two peaks at 660 and 740 keV, due to ^{130}I the isomer ^{130m}I.

The use of an adequate collimator is important in estimating an element in an organ such as the thyroid. The induced activities formed during activation do not all remain at the site of production. Extracellular elements such as sodium and chlorine are removed by the circulating blood and pass out of the field of view of the counter. On the other hand iodine (which is present in the thyroid in organic form) is not subject to this dispersal, despite the possibility of a Szilard-Chalmers reaction during the irradiation.

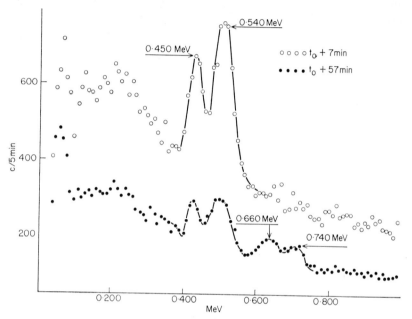

FIG. 11. γ-Ray spectrum from the neck of a sheep irradiated *in vivo* by thermal neutron flux of $3 \cdot 10^8$ n/sq. cm/sec. (Lenihan *et al.*, 1967). Circles—7 min. after irradiation. Crosses—57 min. after irradiation.

The results obtained by Lenihan *et al.* (1967, 1968) were confirmed by activation analysis *in vitro* on the excised thyroid glands after the animals were killed.

Thyroid iodine has been estimated in human subjects by Boddy *et al.* (1968). These workers used a much larger neutron beam (approximately $15 \times 7 \cdot 5$ cm) quite rich in epithermal and fast neutrons. They considered that the irradiation of the thyroid gland was homogeneous and did not use the internal standard technique already described.

Localized irradiation of a superficial bone such as the tibia was used by Comar *et al.* (1968) with the object of estimating the ratio of concentrations of calcium, sodium and chlorine.

The irradiation facility was a curved neutron guide of rectangular cross-section (5 cm × 2 cm) lined internally with a layer of nickel in which neutrons of wavelengths between 0·8 and 40 Å underwent total reflection. In this way the neutron beam emerging from the core of the reactor is, at the exit of the guide, completely free of fast neutrons and γ-radiation. This arrangement allows irradiations to be made at a considerable distance from the reactor and in circumstances combining the advantages of good accessibility with a very small radiation background.

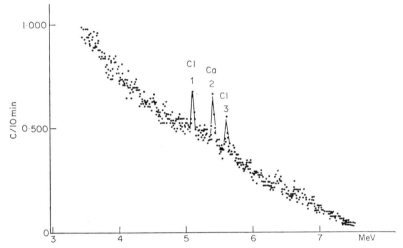

FIG. 12. Spectrum of capture γ-rays obtained during *in vivo* irradiation of the internal anterior aspect of the leg of a normal human subject, using a Ge(Li) detector of volume 20 ml. Peak 1 – 5·090 MeV (chlorine); peak 2 – 5·393 MeV (calcium); peak 3 – 5·598 MeV (chlorine).

The energy of the neutrons emerging from the guide is rather less than the value customarily specified for thermal neutrons and the penetrating power of the beam in tissue is rather small. During irradiation of the inner anterior aspect of the leg with these neutrons only the mineral constituents of the tibia become appreciably radioactive. In these circumstances it is appropriate to study the spectrum of capture γ-rays while the neutron irradiation is in progress. This spectrum shows photo-peaks attributable to chlorine and to calcium (Fig. 12) allowing the ratio of the masses of these two elements in the bone to be estimated.

Repeated measurements of the induced radioactivity in the bone (made after the end of the irradiation, using a suitably collimated scintillation detector) offer the possibility of following the clearance of ^{49}Ca, ^{38}Cl and ^{24}Na formed in the bone under neutron bombardment. It has in this way been possible to confirm that 50% of the bone sodium is exchangeable and that the remaining 50% forms part of the mineral matrix of the bone. The possibility of studying variations in the concentration of these three elements in pathological conditions is of considerable interest, for example in relation to localized decalcification or to callus formation after a fracture.

E. Conclusion

Until quite recently activation analysis was considered as a microanalytical technique finding its main applications in the study of trace elements. Work of this kind is well developed at the present time, thanks to technical improvements including the availability of very high flux reactors, semi-conductor detectors and computer programmes for the analysis of complex γ-ray spectra.

More recently two new areas of application have appeared in the realm of medicine. In the first place, the use of stable activable indicators allows the study of metabolic processes in circumstances where the use of radioactive tracers is impracticable or dangerous – for example in studies on infants or in the investigation of kinetic phenomena extending over a long time. The second significant advance is the development of activation analysis *in vivo*. At the present time this is the only method allowing the direct study of elemental composition of a living organism. For this reason it may prove to have many important clinical applications in the future.

References

Alonzo (d'). C. A., Pell, S. and Fleming, A. J. (1963). *J. occup Med.* **2**, 71.
Anderson, J., Osborn, S. B., Tomlinson, R. W. S., Newton, D., Rundo, J., Salmon, L. and Smith, J. W. (1964). *Lancet ii*, 1201.
Battye, C. K., Tomlinson, R. W. S., Anderson, J. and Osborn S. B. (1967). *In* "Nuclear Activation Techniques in the Life Sciences", p. 573, I.A.E.A. Vienna.
Bergström, J. (1962) *Scand J. clin. Lab. invest.* **14**, suppl. 68.
Bethard, W. F., Olehy, D. A. and Schmitt, R. A. (1964). *In* "L'Analyse par Radioactivation et ses Applications aux Sciences Biologiques" (D. Comar, ed.), pp. 379–393. Presses Universitaires de France, Paris.
Bethard, W. F., Schmitt, R. A., Olehy, D. A., Kaplan, S. A., Ling, S. M., Smith, R. H. and Dalle Molle, E. (1967). *In* "Nuclear Activation Techniques in the Life Sciences", pp. 533–544, I.A.E.A. Vienna.
Biörck, G., Boström, H. and Widström, A. (1965). *Acta med. scand.* **176**, 239.

Boddy, K., Harden, R. McG. and Alexander, W. D. (1968), *J. Clin. Endoc. & Metab.* **28**, 294.
Borg, D. C., Segel, R. E., Kienkle, P. and Campbell, L. (1961). *Int. J. appl. Radiat. Isotopes* **11**, 10.
Bowen, H. J. M. (1956). *J. nucl. Energy* **3**, 18.
Bowen, H. J. M. (1963). Report AERE R-4196.
Bowen, H. J. M. (1966). "Trace Elements in Biochemistry." Academic Press, London.
Bowen, H. J. M. (1967). *Analyst, Lond.* **92**, 124.
Bowen, H. J. M. and Gibbons, D. (1963). "Radioactivation Analysis." Clarendon Press, Oxford.
Brune, D. (1969). *In* "Proceedings of the 1968 International Conference on Modern Trends in Activation Analysis", Gaithersburg, Maryland.
Brune, D. (1966). *Analytica chim. Acta* **34**, 447.
Brune, D. and Landström, O. (1966). *Radiochim. Acta* **5**, 1.
Brune, D., Samsahl, K. and Wester, P. O. (1966). *Clinica chim. Acta* **13**, 285.
Brune, D. and Sjöberg, H. E. (1965). *Analytica chim. Acta* **33**, 570.
Chamberlain, M. J., Fremlin, J. H., Peters, D. K. and Phillip, H. (1968a). *Br. Med. J.* **2**, 581.
Chamberlain, M. J., Fremlin, J. H., Peters, D. K. and Phillip, H. (1968b). *Br. Med. J.* **2**, 583.
Chevallier, F. (1966). *Bull. Soc. Chim. biol.* **48**, 715.
Comar, D., Crouzel, C., Chasteland, M., Rivière, R. and Kellershohn, C. (1969). *Nucl. Appl.* **6**, 344.
Comar, D. and Kellershohn, C. (1961). *In* "Comptes Rendus des Journées d'Études sur l'Analyse par Activation" (J. Laverlochere, ed.), pp. 71–77 C.E.N., Grenoble.
Comar, D., Rivière, R., Raynaud, C. and Kellershohn, C. (1968). *In* "Radioaktive Isotope in Klinik und Forschung, Band VIII". Urban and Schwarzenberg, Munich. [186–192].
Comar, D. and Kellershohn, C. (1967). *In* "Nuclear Activation Techniques in the Life Sciences", pp. 403–417. I.A.E.A., Vienna.
Comar, D. and Le Poec, C. (1965). *In* "Modern Trends in Activation Analysis", pp. 351–356. Activation Analysis Research Laboratory, Texas A and M University, College Station.
Comar, D., Blanquet, P., Rivière, R. and Kellershohn, C. (1963). *In* "Annales de la Faculté des Sciences de l'Université de Clermont-Ferrand" no. 14, 7ème fascicule, Tome V, pp. 23–31. G. de Bussac, Clermont-Ferrand, France.
Comar, D., Rivière, R. and Kellershohn, C. (1964). *In* "L'analyse par radioactivation et ses applications aux sciences biologiques" (D. Comar, ed.), pp. 307–336. Presses Universitaires de France, Paris.
Comar, D., Le Poec, C. and Kellershohn, C. (1967). *In* "Radioaktive Isotope in Klinik und Forschung", Bd VII. Urban and Schwarzenberg, Munich.
Cooper, R. D., Linekin, D. M. and Brownell, G. L. (1967). *In* "Nuclear Activation Techniques in the Life Sciences", pp. 65–80. I.A.E.A., Vienna.
Cotzias, G. C., Papavasiliou, P. S. and Miller, S. T. (1964). *In* "L'analyse par radioactivation et ses applications aux sciences biologiques" (D. Comar, ed.), pp. 287–306. Presses Universitaires de France, Paris.
Cotzias, G. C., Miller, S. T. and Edwards, J. (1966). *J. Lab. clin. Med.* **67**, 836.
De Castro, J. (1949). "Geographie de la faim," Vol. 1. Paris.

Dickson, R. C. and Tomlinson, R. H. (1967). *Int. J. appl. Radiat. Isotopes* **18**, 153.
Dimitriadou, A., Fraser, R. and Turner, P. C. R. (1964). *Nature, Lond.* **201**, 575.
Dubois, J., Colard, J. and Vis, H. L. (1967). In "Nuclear Activation Techniques in the Life Sciences", pp. 545–566. I.A.E.A., Vienna.
Duftschmid, K. E. and Leibetseder, J. (1967). In "Nuclear Activation Techniques in the Life Sciences", pp. 353–363. I.A.E.A., Vienna.
Edwards, R. R. (1962). *Science N.Y.* **137**, 851.
Ekins, R. P. and Sgherzi, A. M. (1965). In "Radiochemical Methods of Analysis", Vol. II, 235, I.A.E.A., Vienna.
Fireman, P., Borg, D. and Gitlin, D. (1966). *Nature, Lond.* **210**, 547.
Fite, L. E., Wainerdi, R. E., Yule, H. P., Harrison, G. M., Bickers, G. and Doggett, R. (1969). In "Proceedings of the 1968 International Conference on Modern Trends in Activation Analysis", Gaithersburg, Maryland.
Fleckenstein, A., Gerlach, E., Janke, J. and Marmier, P. (1960). *Pflügers Arch. ges. Physiol.* **271**, 75.
Fleischer, A. A. (1965). Report no 2001, The Cyclotron Corp., Berkeley.
Fukai, R. (1959). Report AECU-3887.
Furst, A. (1960). In "Metal Binding in Medicine" (M. J. Seven, ed), pp. 336–344. J. B. Lippincott, Philadelphia.
Girardi, F. (1969). In "Proceedings of the 1968 International Conference on Modern Trends in Activation Analysis", Gaithersburg, Maryland.
Girardi, F., Merlini, M., Pauly, J. and Pietra, R. (1964). In "Symposium on Radiochemical Methods of Analysis". I.A.E.A., Vienna.
Gleit, C. E., Benson, P. A., Holland, W. D. and Russel, I. J. (1962). *Analyt. chem.* **34**, 1454.
Hanson, A. and Biörck, G. (1957). *Acta med. scand.* **157**, 493.
Herring, W. B., Leavell, B. S., Paixao, L. M. and Yoe, J. H. (1960). *Am. J. clin. Nutr.* **8**, 846.
Horwitt, M. K. (1965). *Fedn Proc. Fedn Am. Socs exp. Biol.* **24**, 68.
Javillier, M., Polonowski, M. and Florkin, M. (1959). "Traité de biochimie générale", Tome 1, deuxième fascicule. Masson et Cie, Paris.
Kanabrocki, E. L., Fields, T., Decker, C. F., Case, L. F., Miller, E. B., Kaplan, E. and Oester, Y. T. (1964). *Int. J. appl. Radiat. Isotopes* **15**, 175.
Kehoe, R. A., Cholak, I. and Stary, R. V. (1940). *J. Nutr.* **19**, 579.
Keisch, B., Koch, R. C. and Levine, A. S. (1965). In "Modern Trends in Activation Analysis", pp. 284–290. Activation Analysis Research Laboratory, Texas A and M University, College Station.
Kellershohn, C., Comar, D. and Le Poec, C. (1965). *J. Lab. clin. Med.* **66**, 168.
Kellershohn, C., Comar, D. and Rivière, R. (1967). *C.r. hebd Séanc. Acad. Sci., Paris, B*, **264**, 1836, 1967 and 265, 88, 1967.
Linekin, D. M., Balcius, J. F., Cooper, R. D. and Brownell, G. L. (1969). In "Proceedings of the 1968 International Conference on Modern Trends in Activation Analysis", Gaithersburg, Maryland.
Lenihan, J. M. A., Comar, D., Rivière, R. and Kellershohn, C. (1968). *J. Nuc. Med.* **9**, 110.
Lenihan, J. M. A. and Thomson, S. J. (1965). "Activation Analysis." Academic Press, London.
Lenihan, J. M. A., Comar, D., Rivière, R. and Kellershohn, C. (1967). *Nature Lond.* **214**, 1221.

Lissitzky, S., Comar, D., Rivière, R. and Codaccioni, J. L. (1965). *Rev. Fse Et. Clin. Biol.* **10**, 631.
Livingstone, H. B. and Smith, H. (1967). *Analyt. chem.* **39**, 538.
Lowman, J. T. and Krivit, W. (1967). In "Nuclear Activation Techniques in the Life Sciences". I.A.E.A., Vienna.
Maletskos, C. J., Keane, A. T., Littlefields, S. and McClatchey, R. A. (1961). M.I.T. Radioactivity Center. *Annual Progress Report Document*, **952**, 68.
Maletskos, C. J., Osborn, H. F. and Braun, A. G. (1963). M.I.T. Radioactivity Center. *Annual Progress Report Document*, **9505**, 47.
Manney, T. R., La Roche, G., Padden, R. E. and Tobias, C. A. (1961). *Report U.C.R.L.* 9897.
McPherson, D. (1965). *Acta orthop. scand.* Suppl. **78**
Meinke, W. W. (1963). In "Chemistry Research and Chemical Techniques Based on Research Reactors". pp. 73–82. I.A.E.A., Vienna.
Mulvey, P. F., Cardorelli, J. A., Meyer, R. A., Cooper, R. and Burrows, B. A. (1965). In "Radioisotope Sample Measurement Techniques in Medicine and Biology", pp. 249–258. I.A.E.A., Vienna.
Nagaï, T., Fujii, I. and Muto, H. (1968). *J. nucl. Med.* **9**, 338.
Newton, D., Anderson, J., Battye, C. K., Osborn, S. B. and Tomlinson, R. W. S. (1969). *Int. J. appl. Radiat. Isotopes*, **20**, 61.
Odeblad, E. (1956). *Clin. chim. Acta* **1**, 67.
Orr, J. S. and Gillespie, F. C. (1968). *Science, N.Y.* **162**, 138.
Ouellette, R. P., Balcius, J. F. and Zuppinger, K. (1966). *Int. J. appl. Radiat. Isotopes* **17**, 649.
Palmer, H. E., Nelp, W. B., Murano, R. and Rich, C. (1968). *Phys. Med. Biol.* **13**, 269.
Parr, R. M. and Taylor, D. M. (1964). *Biochem. J.* **91**, 424.
Peisach, M. and Pretorius, R. (1966). *Analyt. chem.* **38**, 956.
"Radiochemical Methods of Analysis" (1965). I.A.E.A. Vienna.
Rappaport, R., Comar, D., Rivière, R. and Royer, P. (1967). *Arch. Franç. Ped.* (Sem. Hôp.). **24**, 399–413.
Renzo (de) E. C. (1962). In "Mineral Metabolism" (C. L. Comar and F. Bronner, eds), Vol. 2, part B, pp. 483–498. Academic Press, New York.
Rivière, R. (1968). Private communication.
Rivière, R., Comar, D. and Kellershohn, C. (1965). In "Current Topics in Thyroid Research", pp. 112–121, Proceedings of the Fifth Int. Conf. Thyroid., Academic Press, New York.
Rivière, R., Comar, D., Kellershohn, C., Mandry, J., Juif, J. G., Sacrez, R. and Methlin, G. (1967). *Annls. pediat.* (Sem. Hôp.) **43**, 10.
Rivière, R., Comar, D., Kellershohn, C., Orr, J. S., Gillespie, F. C. and Lenihan, J. M. A. (1969), *Lancet i*, 339.
Rosenfeld, T. and Beath, O. A. (1964). "Selenium." Academic Press, London.
Rundo, J. and Bunce, L. J. (1966). *Nature, Lond.* **210**, 1023.
Ruzicka, J. and Stary, J. (1964). In "Atomic Energy Rev". **2**, 3. I.A.E.A., Vienna.
Samsahl, K. (1966). *Report A.E.* 247, Aktiebolaget. Atomenergi, Stockholm.
Samsahl, K. and Brune, D. (1965). *Int. J. appl. Radiat. Isotopes* **16**, 273.
Samsahl, K., Brune, D. and Wester, P. O. (1963). *Report A.E.*—124, Aktiebolaget. Atomenergi Stockholm.
Schroeder, H. A. (1960). In "Metal Binding in Medicine" (M. J. Seven, ed.), pp. 59–67. J. B. Lippincott, Philadelphia.

Simon, C. (1964). "Une nouvelle méthode d'étude du métabolisme de l'iode; la méthode d'équilibre isotopique. Aspects cinétiques et quantitatifs obtenus chez le rat". Thèse de sciences, Paris.

Sklavenitis, H. and Comar, D. (1967). *In* "Nuclear Activation Techniques in the Life Sciences", pp. 435–444. I.A.E.A., Vienna.

Sklavenitis, H., Devillers, C., Comar, D. and Kellershohn, C. (1969). *Int. J. appl. Radiat. Isotopes.* (In the press.)

Smith, E. M., Mozley, J. M. and Wagner, H. H. (1964). *J. nucl. Med.* **5**, 828.

Smith, H. (1963). *Analyt. chem.* **35**, 749.

Smith, J. W. (1962). *Physics. Med. Biol.* **7**, 341.

Soremark, R. (1960). *Acta physiol. scand.* **50**, 306.

Spector, W. S. (1956). "Handbook of Biological Data." W. B. Saunders Company, London.

Stary, J. and Ruzicka, J. (1964). *Talanta* **11**, 697.

Strain, W. H. (1961). "Effects of Some Minor Elements on Animals and People." Proceedings of a Symposium "Geochemical Evolution—The First Five Billion Years." Am. Ass. for the Advancement of Science, Denver, Colorado.

Stubbins, M. I. and Fremlin, J. H. (1967). *In* "Nuclear Activation Techniques in the Life Sciences", pp. 463–478. I.A.E.A., Vienna.

Tang, C. W. and Tomlinson, R. H. (1967). *In* "Nuclear Activation Techniques in the Life Sciences", pp. 427–434. I.A.E.A., Vienna.

Teitz, N. W., Hirsch, E. F. and Neyman, B. (1957). *J. Am. med. Ass.* **165**, 1.

Tipton, I. H. (1960). *In* "Metal Binding in Medicine" (M. J. Seven, ed.), pp. 27–42. J. B. Lippincott, Philadelphia.

Tomlinson, R. H. and Dickson, R. C. (1965). *In* "Modern Trends in Activation Analysis", pp. 66–71. Activation Analysis Research Laboratory, Texas A and M University, College Station.

Vartapetyan, B. B., Dmitrovskii, A. A., Alkhazov, D. G., Lemberg, I. Kh., Girshin, A. B., Gusinskii, G. M., Starikova, N. A., Erofeeva, N. N., and Bogdanova, I. P. (1966). *Biochemistry*, **31**, 5.

Wacker, W. E. C., Adelstein, S. J., Ulmer, D. D. and Vallee, B. L. (1956). *J. clin. Invest.* **35**, 741.

Wagner, H. H., Nelp, W. B. and Dawling, J. H. (1961). *J. clin. Invest.* **40**, 1984.

Wahl, W. H., Nass, H. W. and Kramer, H. H. (1966). *In* "Radioactive Pharmaceuticals" (G. A. Andrews, R. M. Kniseley, H. N. Wagner, eds), 191. Distributed by Clearinghouse for Federal Scientific and Technical Information, Springfield, Virginia, U.S.A.

Warburg, O. (1956). *Science N.Y.* **123**, 309.

Wester, P. O. (1965). *Acta med. scand.* suppl. 439.

Widdowson, E. M. and Dickerson, J. W. T. (1964). *In* "Mineral Metabolism" (C. L. Comar and F. Bronner, eds), Vol. 2, part A, pp. 1–247. Academic Press, New York.

Woodruff, G. L., Wilson, W. E., Yamamoto, Jn. Y., Babb, A. L. and Stamm, S. J. (1969). *In* "Proceedings of the 1968 International Conference on Modern Trends in Activation Analysis", Gaithersburg, Maryland.

AUTHOR INDEX

The numbers in *italics* indicate pages on which names are mentioned in the reference lists

A

Adams, F., 148, *160*
Adelstein, S. J., 179, *206*
Alexander, W. D., 195, 200, *203*
Alkhazov, D. G., 172, *206*
Alsmiller, R., 116, *134*
Anderson, J., 172, 194, 195, 196, 197, *202, 205*

B

Babb, A. L., 183, *206*
Baccius, J. F., 176, *204*
Balcius, J. F., 192, *205*
Battye, C. K., 196, *202, 205*
Beath, O. A., 192, *205*
Becker, D. A., 160, *161*
Benson, P. A., 169, *204*
Bergström, T., 183, *202*
Bertini, H., 116, *134*
Bertolini, G., 138, *160*
Bethard, W. F., 166, 168, 188, *202*
Bickers, G., 183, *204*
Bilger, H. R., 141, *161*
Biörck, G., 179, *202, 204*
Bivins, R., 116, *134, 135*
Black, J. L., 149, *160*
Black, W. W., 138, 141, 142, 146, 147, *160, 161*
Blanquet, P., 169, 182, *203*
Boddy, K., 195, 200, *203*
Bogdanova, I. P., 172, *206*
Borella, A., 152, 153, *161*
Borg, D. C., 171, *203, 204*
Boström, H., 179, *202*
Bowen, H. J. M., 105, 106, *113*, 164, 165, 166, 168, 170, 174, *203*
Branstetter, E., 125, *135*
Braun, A. G., 193, *205*

Breen, W. M., 98, *99*
Brish, J., 116, 134
Brown, R., 103, *113*
Brownell, G. L., 155, *160*, 172, 176, *203, 204*
Brune, D., 165, 170, 173, 174, 175, 179, *203, 205*
Bruninx, E., 117, 119, 123, *134*
Bryan, D. B., 157, *160*
Buchanan, J. D., 59, 61, *78*
Bunce, L. J., 198, *205*
Burrows, B. A., 172, 180, *205*

C

Campbell, L., 171, *203*
Cappellani, F., 143, *160*
Cardorelli, J. A., 172, 180, *205*
Case, L. F., 166, 169, 179, *204*
Cazier, G. A., 97, *100*, 148, *161*
Chamberlain, M. J., 195, 197, 198, *203*
Charalambus, S., 123, 125, *134*
Chasteland, M., 177, *203*
Chevallier, F., 188, *203*
Cholak, I., 166, *204*
Chupp, E., 116, *134*
Cline, J. E., 138, 141, 142, 146, *161*
Cobb, J. C., 156, *160*
Coche, A., 138, *160*
Codaccioni, J. L., 182, *205*
Colard, J., 183, *204*
Comar, D., 164, 167, 168, 169, 170, 173, 174, 177, 180, 181, 182, 184, 185, 186, 190, 191, 192, 193, 195, 199, 200, *203, 205, 206*
Cook, G. B., 101, 104, *113*
Cooper, J. A., 156, 157, 160, *160*
Cooper, R. D., 155, *160*, 172, 176, 180, *203, 204, 205*

AUTHOR INDEX

Cotzias, G. C., 166, 167, 168, 169, 187, 203
Covell, D. F., 82, *99*
Crespi, M. B. A., 101, 104, *113*
Crouzel, C., 177, *203*
Cumming, J., 118, 119, 123, *134*, *135*
Cuypers, M. Y., 99, *100*

D

d'Alonzo, C. A., 179, *202*
Dalle Molle, E., 188, *202*
Dams, R., 148, *160*
Dawling, J. H., 180, *206*
Day, R. B., 149, 150, *161*
Dearnaley, G., 138, *160*
De Castro, J., 178, *203*
Decastro Faria, N. V., 149, *160*
Decker, C. F., 166, 169, 179, *204*
De Hevesy, G., 81, *99*
Devillers, C., 193, 195, *206*
Dickerson, J. W. T., 165, *206*
Dickson, R. C., 167, 177, 182, *204*, *206*
Dimitriadou, A., 192, *204*
Dmitrovskii, A. A., 172, *206*
Dogget, R., 183, *204*
Dorpema, B., 148, *161*
Dostrovsky, I., 116, *134*
Dubois, J., 183, *204*
Duftschmid, K. E., 182, *204*
Dutrannois, J., 123, 125, *134*

E

Edwards, J., 166, 167, 168, 187, *203*
Edwards, R. R., 190, *204*
Eeckhaut, Z., 32, *36*
Ekins, R. P., 167, *204*
Engelmann, C., 30, 31, 35
Erofeeva, N. N., 172, *206*
Erwall, L. G., 78, *78*
Evans, R. D., 155, 156, 157, *161*
Ewan, F. T., 141, 150, *160*, *161*

F

Fairbairn, H. W., 101, 102, 103, *113*
Fields, T., 166, 169, 179, *204*
Fireman, P., 192, *204*
Fite, L. E., 89, 93, 98, 99, *99*, *100*, 183, *204*
Flanagan, F. J., 102, 104, *113*
Fleckenstein, A., 172, *204*
Fleischer, A. A., 172, *204*
Fleischer, M., 102, 103, *113*
Fleming, A. J., 179, *202*
Florkin, M., 165, *204*
Fowler, I. L., 141, *160*
Fraser, R., 192, *204*
Freeman, J. M., 149, 152, *161*
Fremlin, J. H., 172, 195, 197, 198, *203*, *206*
Friedlander, G., 116, 118, 135
Fry, E. S., 149, 150, *161*
Fujii, I., 198, *205*
Fukai, R., 173, *204*
Fumagalli, W., 143, *160*
Furst, A., 179, *204*

G

Gerlach, E., 172, *204*
Gibbons, D., 89, 93, 98, 99, *99*, *100*, 168, 174, *203*
Gillespie, F. C., 190, *205*
Girardi, F., 142, 146, 147, 148, 149, 150, 151, 152, 153, 154, 155, 158, 159, 160, *161*, 175, *204*
Girshin, A. B., 172, *206*
Gitlin, D., 192, *204*
Gleit, C. E., 169, *204*
Goebel, K., 123, 125, *134*
Groves, D. J., 146, *161*
Grühle, W., 149, *160*
Guinn, V. P., 5, *35*, 73, *78*, 157, *160*
Gusinskii, G. M., 172, *206*
Guzzi, G., 142, 146, 148, 149, 150, 152, 153, 154, 155, 158, 159, *161*
Gwyn, M. E., 104, *113*

H

Hahn, O., 38
Hahn, R. L., 31, *36*
Haller, W. A., 160, *161*
Hanson, A., 179, *204*
Harden, R. McG., 195, 200, *203*
Harris, J. A., 152, 155, *161*
Harrison, G. M., 183, *204*
Hawton, J. J., 57, 59, *78*
Heath, R. L., 57, *78*, 97, *100*, 138, 141, 142, 146, 147, 148, *160*, *161*
Heeney, H. B., 101, *113*
Helmer, R. G., 97, *100*, 148, *161*
Henuset, M, 143, *160*
Herring, W. B., 165, *204*

Hess, W., 116, *134*
Hines, J., 34, *36*
Hirsch, E. F., 178, 179, *206*
Holland, W. D., 169, *204*
Hollander, J. M., 57, *78*, 152, 155, *161*
Horwitt, M. K., 182, *204*
Hoste, J., 32, *36*
Hotz, H. P., 149, *161*
Howerton, R. J., 29, *35*
Hudis, J., 116, 118, *135*
Hughes, D. J., 29, *35*, 57, *78*
Hurley, J. P., 149, *161*

J

Janke, J., 172, *204*
Javillier, M., 165, *204*
Jenkin, J. G., 149, 152, *161*
Jervis, R. E., 160, *161*
Juif, J. G., 182, *205*

K

Kanabrocki, E. L., 166, 169, 179, *204*
Kaplan, E., 166, 169, 179, 188, *202*, *204*
Kaufman, S., 118, *135*
Keane, A. T., 193, *205*
Kehoe, R. A., 166, *204*
Keisch, B., 177, *204*
Kellershohn, C., 164, 167, 168, 169, 170, 173, 177, 180, 181, 182, 184, 185, 190, 191, 193, 195, 199, 200, *203*, *204*, *205*, *206*
Kenworthy, A. L., 101, *113*
Kienkle, P., 171, *203*
Kinnen, W. E., 125, *135*
Knoll, G. F., 149, *161*
Koch, R. C., 177, *204*
Kolbe, P., 104, *113*
Kramer, H. H., 188, *206*
Krivit, W., 190, *205*
Kuykendall, W. E. Jr., 82, *100*

L

Lamb, J. F., 152, 155, *161*
Landström, O., 170, *203*
La Roche, G., 180, *205*
Leavell, B. S., 165, *204*
Leibetseder J., 182, *204*
Leimdorfer, M., 116, *134*
Leliaert, G., 32, *36*

Lemberg, I. Kh., 172, *206*
Lenihan, J. M. A., 174, 190, 195, 199, 200, *204*, *205*
Le Poec, C., 164, 168, 173, 174, 182, 185, *203*, *204*
Levesque, R. J. A., 149, *160*
Levi, H., 81, *99*
Levine, A. S., 177, *204*
Linekin, D. M., 155, *160*, 172, 176, *203*, *204*
Ling, S. M., 188, *202*
Liskien, H., 121, *135*
Lissitzky, S., 182, *205*
Littlefields, S., 193, *205*
Livingstone, H. B., 183, *205*
Lowman, J. T., 190, *205*
Lukens, H. R., 31, *36*, 57, 61, 64, *78*
Lyon, W. S. Jr., 30, *36*, 157, *161*

M

McCaslin, J., 125, 126, 127, *135*
McClatchey, R. A., 193, *205*
McMahon, J., 82, *100*
McPherson, D., 189, *205*
Mahony, J. D., 30, *36*
Maletskos, C. J., 193, *205*
Malm, H. L., 141, *160*
Mandry, J., 182, *205*
Mann, H. M., 141, *161*
Manney, T. R., 180, *205*
Markowitz, S. S., 30, 36
Marmier, P., 172, *204*
Mathiensen, J. M., 149, *161*
Mathis, W. T., 101, *113*
Meinke, W. W., 174, *205*
Menon, M. P., 89, 99, *100*, 160, *161*
Merlini, M., 160, *161*, 175, *204*
Methlin, G., 182, *205*
Metropolis, N., 116, *135*
Meyer, R. A., 172, 180, *205*
Miller, E. B., 166, 169, 179, *204*
Miller, E. J., 101, *113*
Miller, J., 116, *135*
Miller, S. T., 166, 167, 168, 169, 187, *203*
Minczewski, J., 101, 104, *113*
Moyer, B., 116, *135*
Mozley, J. M., 180, *206*
Mulvey, P. F., 172, 180, *205*
Murano, R., 195, 196, 197, *205*
Muto, H., 198, *205*

N

Nagai, T., 198, *205*
Nass, H. W., 188, *206*
Nelp, W. B., 180, 195, 196, 197, *205, 206*
Neufeld, J., 125, *135*
Newton, D., 172, 194, 195, 196, 197, *202, 205*
Neyman, B., 178, 179, *206*
Northrop, D. C., 138, *160*

O

O'Barrell, M., 116, *135*
Odeblad, E., 188, *205*
Oester, Y. T., 166, 169, 179, *204*
Olehy, D. A., 166, 168, 188, *202*
Organesian, K., 121, *135*
Orr, J. S., 190, *205*
Osborn, H. F., 193, *205*
Osborn, S. B., 172, 194, 195, 196, 197, *202, 205*
Otvos, J. W., 31.
Ouellette, R. P., 192, *205*

P

Padden, R. E., 180, *205*
Paixao, L. M., 165, *204*
Palmer, H. E., 156, 157, 160, *160*, 195, 196, 197, *205*
Palms, J. M., 149, 150, *161*
Panofsky, W., 119, *135*
Papavasiliou, P. S., 169, *203*
Parr, R. M., 165, 167, *205*
Patterson, H., 116, *134*
Patterson, W., 125, 126, 127, *135*
Paulsen, A., 121, *135*
Pauly, J., 142, 146, 147, 148, 149, 150, 151, 152, 153, 154, 155, 158, 159, 160, *161*, 175, *204*
Peisach, M., 172, *205*
Pell, S., 179, *202*
Perkins, R. W., 155, 156, 157, 160, *160, 161*
Peters, D. K., 195, 197, 198, *203*
Phillip, H., 195, 197, 198, *203*
Phillips, R., 119, *135*
Pietra, R., 175, *204*
Polonowski, M., 165, *204*
Poskanzer, A., 118, *135*
Pozzi, G., 160, *161*

Pretorius, R., 172, *205*
Prussin, S J., 152, 155, *161*

R

Rabinowitz, P., 116, *134*
Radin, N., 105, *113*
Ragaini, R. G., 155, 156, 157, *161*
Rainosek, A. P., 99, *100*
Rappaport, R., 182, *205*
Raynaud, C., 195, 200, *203*
Reeder, P. L., 118, *135*
Renzo, de E. C., 183, *205*
Restalli, G., 143, *160*
Ricci, E., 31, *36*
Rich, C., 195, 196, 197, *205*
Rivière, R., 169, 177, 180, 181, 18 2, 184 185, 190, 192, 195, 199, 200, *203, 205*
Robertson, D. E., 155, *161*
Rosenfeld, A., 121, *135*
Rosenfeld, T., 182, *205*
Rossi, B., 116, *135*
Roubault, M., 102, *113*
Roy, J. C., 57, 59, *78*
Royer, P., 182, *205*
Rubinson, W., 7, *36,*
Rundo, J., 172, 194, 195, 197, 198, *202, 205*
Russel, I. J., 169, *204*
Ruzicka, J., 175, *205, 206*

S

Sabbioni, E., 142, 146, 147, 151, 160, *161*
Sacrez, R., 182, *205*
Salmon, L., 82, *100*, 172, 194, 195, 197, *202*
Samsahl, K., 165, 175, 179, *203, 205*
Schairer, J. F., 101, *113*
Scherzi, A. M., 167, *204*
Schmitt, R. A., 166, 168, 188, *202*
Schmittroth, L. A., 97, *100*, 148, *161*
Schroeder, G. L., 155, 156, 157, *161*
Schroeder, H. A., 165, 179, *205*
Schulze, W., 12, *26*, 34, *36*, 148, *161*
Schumann, R., 82, *100*
Schwartz, R. B., 29, *35*, 57, *78*
Seaborg, G. T., 57, *78*
Segel, R. E., 171, *203*
Settle, D. M., 157, *160*
Shermann, I. S., 141, *161*
Simon, C., 191, 192, *206*

Sjöberg, H. E., 174, *203*
Sklavenitis, H., 185, 186, 192, 193, 195, *206*
Sklavenitis, L., 121, 123, 127, 130, *135*
Smith, A., 125, 126, 127, *135*
Smith, E. M., 180, *206*
Smith, G. W., 160, *161*
Smith, H., 183, 189, *205, 206*
Smith, J. W., 172, 194, 195, 196, 197, *202, 206*
Smith, L. H., 99, *100*
Smith, R. H., 188, *202*
Snyder, W., 125, *135*
Sondhaus, C., 116, *135*
Soremark, R., 187, *206*
Specker, H., 35, *36*
Spector, W. S., 165, *206*
Stamm, S, J., 183, *206*
Stary, J., 175, *205, 206*
Stary, R. V., 166, *204*
Steele, E. L., 30, *36*
Stephens, L., 125, 126, 127, *135*
Storm, M., 116, *135*
Straikova, N. A., 172, *206*
Strain, W. H., 165, 178, *206*
Strassman, F., 38
Strominger, D., 57, *78*
Stubbins, M. I., 172, *206*
Swanson, R., 121, *135*

T

Tang, C. W., 160, *161*, 176, *206*
Tavendale, A. J., 141, 150, *161*
Taylor, D. M., 165, 167, *205*
Taylor, J. M., 138, *161*
Taylor, S. R., 104, *113*
Teitz, N. W., 178, 179, *206*
Thomson, S. J., 174, *204*
Tipton, I. H., 165, *206*
Tobias, C. A., 180, *205*
Tomlinson, C. W., 160, *161*
Tomlinson, R. H., 167, 176, 177, 182, *204, 206*
Tomlinson, R. W. S., 172, 194, 195, 196, 197, *202, 205*
Torpe, J. D., 160, *161*
Turkevick, A., 116, *135*
Turner, J., 125, *135*
Turner, J. E., 125, *135*
Turner, P. C. R., 192, *204*

U

Ulmer, D. D., 179, *206*

V

Vallee, B. L., 179, *206*
Van Wyk, J. M., 99, *100*
Vartapetyan, B. B., 172, *206*
Vis, H. L., 183, *204*
Vos, G., 142, 146, 147, 150, 160, *161*

W

Wacker, W. E. C., 179, *206*
Wagner, C. D., 31,
Wagner, H. H., 180, *206*
Wahl, W. H., 188, *206*
Wahlgren, M., 34, *36*
Wainerdi, R. E., 82, 89, 93, 98, 99, *99*, *100*, 160, *161*, 183, *204*
Waino, K. M., 149, *161*
Wallace, R., 116, *134, 135*
Warburg, O., 179, *206*
Ward, G. M., 101, *113*
Warshaw, S., 121, *135*
Watt, B. E., 39, *78*
Wester, P. O., 165, 175, 179, *203, 205, 206*
White, D. H., 146, *161*
Widdowson, E. M., 165, *206*
Widström, A., 179, *202*
Wilson, W. E., 183, *206*
Wing, J., 34, *36*
Wogman, N. A., 156, 157, 160, *160*
Wolstenholme, W. A., 103, *113*
Wong, K. Y., 160, *161*
Woodruff, G. L., 183, *206*
Woodyard, R. L., 125, *135*
Wright, H. A., 125, *135*
Wyckoff, T. M., 148

Y

Yamamoto, Tn. Y., 183, 206
Yoe, J. H., 165, *204*
Yule, H. P., 73, *78*, 96, *100*, 148, *161*, 183, *204*

Z

Zerby, C. D., 125, *135*
Zuppinger, K., 192, *205*

SUBJECT INDEX

A

^{41}A, 158
Accelerator, 10, 86, 125
Activation, 4, 10, 31, 86, 88
   by α-particles, 30
   by charged particles, 30
   by deuterons, 30
   by fast neutrons, 10, 60
   by γ-rays, 3
   by ^{3}He particles, 30, 31
   by neutrons, 4
   by nuclei, 4
   by photons, 31
   by protons, 30
   by thermal neutrons, 4
   for high energy particle fluence assessment, 118, 121
   irradiation period, 64, 65, 66
   methods, 3
   producing short-lived radionuclides, 5
   recoil protons, 68
Activation analysis
   accuracy and precision, 101, 103, 104
   advantages, 2, 35
   automation, 82, 83, 84, 89, 90, 94
   biological standards, 105, 106
   chemical treatment of sample, 2
   comparison with other methods, 35
   contamination in, 35, 82, 82, 91
   data handling, 94, 95, 96, 97, 98, 99: by computer, 82, 89, 94, 95, 99
   in biology, 157, 160
   in cosmochemistry, 103
   in fluence assessment, 115, 127-130
   in forensic science, 157
   in geochemistry, 102, 103, 155, 155
   in historical studies, 157

   in vivo, 71, 192-202
   instrumentation, 82
   intercomparison, 101, 102, 104, 106
   kale standard, 106, 107, 108, 109, 110, 111
   limitations, 2, 30
   of high-purity materials, 155
   packaging of sample, 89, 91
   rock standards, 102, 103, 104
   sample irradiation, 62, 63, 64, 66, 88
   sensitivity, 30, 35
   standards, 101, 102, 104, 105, 106
   storage of sample, 91, 92
Activation analysis in clinical science, 164, 166, 167, 168, 177-202
   activity measurement, 176, 177
   automation, 174, 175
   blood, 170, 171, 176, 180, 182, 184, 187, 189, 192
   bone, 172, 189, 192, 193, 197, 200, 201, 202
   bromine metabolism, 185, 186, 192
   calcium metabolism, 188, 189
   choice of tracer, 188, 189, 190, 192, 193
   carcinogenicity, 178, 179
   contamination, 166, 169, 173
   coronary disease, 179, 180
   dental enamel, 172
   dose, 193, 194
   iodine metabolism, 190
   iron metabolism, 189
   liver, 173
   manganese metabolism, 187, 188
   metabolic studies, 183-192, 202
   muscle, 173
   plasma, 166, 175, 179, 180, 181, 182, 185, 186, 190, 191, 192
   proteins, 192

Activation analysis in clinical science, (*contd.*)
  radiation sources, 171, 172
  red cells, 172, 182, 189
  sample irradiation, 170, 171
  sample preparation, 169, 172
  sample treatment, 173
  serum, 173, 186
  standardization, 170
  thyroid, 181, 182, 184, 185, 190, 191, 192, 193, 199, 200
  tissue, 174, 189
  total body calcium, 197
  total body sodium, 196, 197
  tumour tissue, 179
  urine, 175, 180, 191
  Vitamin A in rat, 172
  whole body, 194-8
Activation detector, 89, 118, 125, 127, 130, 133
  advantages, 133
  as particle dosimeter, 125, 131
  for fluence assessment, 118, 125, 127, 133, 134
^{105}Ag, 20
^{106}Ag, 20
^{107}Ag, 20
^{108}Ag, 19, 20
^{109}Ag, 19, 20, 21
^{110}Ag, 19, 20, 21, 155, 171
^{111}Ag, 19, 20
^{112}Ag, 20, 21
^{113}Ag, 20
^{114}Ag, 20
^{116}Ag, 20
^{24}Al, 129
^{26}Al, 13
^{27}Al, 11, 29, 55, 117, 119, 120, 121, 122, 123, 128, 129
^{28}Al, 11, 13, 29, 52, 158, 194, 196, 198
  as flux monitor, 52
  γ-ray energy, 52
^{29}Al, 13
^{241}Am, 147
^{35}Ar, 14
^{37}Ar, 14
^{38}Ar, 57
^{39}Ar, 14
^{40}Ar
^{41}Ar, 14, 45, 66
^{43}Ar, 14
^{45}Ar, 14
^{47}Ar, 14
Arsenic
  assay, 32
^{73}As, 16
^{74}As, 16
^{75}As, 75
^{76}As, 16, 17
^{77}As, 16
^{78}As, 16, 17
^{79}As, 16
^{80}As, 16
^{82}As, 16
^{195}Au, 29
^{196}Au, 28, 29
^{197}Au, 28, 75, 117, 120, 123
^{198}Au, 28, 29, 53, 147
^{199}Au, 28, 29
^{200}Au, 29
^{201}Au, 29
^{202}Au, 29
^{204}Au, 28

B

^{10}B, 42, 171, 172
^{11}B, 42, 76, 120, 121, 123, 129, 130
^{129}Ba, 23
^{131}Ba, 23, 24, 156
^{133}Ba, 23, 24
^{135}Ba, 23, 24
^{136}Ba, 58
^{137}Ba, 23, 24, 58
^{138}Ba, 58
^{139}Ba, 23, 24
^{140}Ba, 157
Background count, 34
  effect on limits of detection, 34
^{7}Be, 119, 123, 126
^{9}Be, 41
  as reactor moderator, 41, 42, 44, 51
  diffusion length, 43
  Fermi age, 42
  moderating ratio, 42
  slowing-down length, 43, 51
^{10}Be, 13
^{11}Be, 13
^{207}Bi, 147
^{208}Bi, 29
^{210}Bi, 29

SUBJECT INDEX

^{212}Bi, 150
Blood
    elemental composition, 165
^{77}Br, 17
^{78}Br, 16, 17
^{80}Br, 16, 17, 175
^{82}Br, 16, 17, 159, 171, 173, 185, 192
^{83}Br, 16, 17
^{84}Br, 17
^{86}Br, 17
β-radiation, 45

C

^{11}C, 13, 117, 118, 119, 121, 123, 126, 129, 130, 132
^{12}C, 41, 71, 117, 118, 119, 120, 121, 122, 123, 126, 129, 132
    as reactor moderator, 41, 42
    diffusion length, 43
    Fermi age, 42
    moderating ratio, 42
    slowing-down length, 43
^{13}C, 69, 189
^{14}C, 13, 196
^{15}C, 13, 45
^{39}Ca, 14
^{41}Ca, 14
^{44}Ca, 172
^{45}Ca, 14
^{46}Ca, 189
^{47}Ca, 14, 157, 159, 173, 189
^{48}Ca, 172, 189
^{49}Ca, 14, 158, 189, 194, 196, 197, 202
^{105}Cd, 20
^{107}Cd, 20
^{109}Cd, 20, 21
^{111}Cd, 19, 20, 21
^{113}Cd, 6, 20, 21
^{114}Cd, 6
^{115}Cd, 20, 21
^{117}Cd, 20, 21
^{119}Cd, 21
^{121}Cd, 21
^{135}Ce, 24
^{137}Ce, 24
^{139}Ce, 24
^{141}Ce, 24, 156
^{143}Ce, 24
^{145}Ce, 24
^{147}Ce, 24
Cerenkov radiation, 47
^{34}Cl, 14
^{35}Cl, 54, 56
^{36}Cl, 14
^{37}Cl, 57
^{38}Cl, 14, 55, 57, 150, 158, 173, 194, 196, 197, 202
^{40}Cl, 14
^{57}Co, 15
^{58}Co, 15
^{59}Co, 55, 125
^{60}Co, 15, 85, 86, 125, 142, 147, 150, 156, 157, 159, 160, 171, 173
^{61}Co, 15
^{62}Co, 15
^{64}Co, 15
^{66}Co, 53
Cockroft-Walton generator
    in clinical activation analysis, 195
Coincidence, 33
    in analysis, 34
Compton scattering, 155
Control rods, 46
^{49}Cr, 15
^{50}Cr, 189
^{51}Cr, 15, 156, 173, 189
^{52}Cr, 75
^{54}Cr, 55, 58
^{55}Cr, 15
Cross-section 7, 10, 29, 45, 47, 48, 55, 56, 57, 58, 62, 63, 64, 71, 76, 117, 118, 119, 121, 123, 126, 128, 132 energy curves, 123, 132
    for ^{235}Cl, 39, 40
^{129}Cs, 23
^{130}Cs, 23
^{131}Cs, 23, 24
^{132}Cs, 23
^{134}Cs, 23, 24, 156, 157, 173
^{135}Cs, 24
^{136}Cs, 24
^{137}Cs, 23, 24, 142, 146, 147, 150, 196
^{138}Cs, 23
^{62}Cu, 15
^{64}Cu, 15, 52, 171, 173
    as flux monitor, 52
    γ-ray energy, 52
^{66}Cu, 15, 16
^{67}Cu, 15

^{68}Cu, 15
^{70}Cu, 15
Cyclotron, 86, 172
   in clinical activation analysis, 172, 195, 196

D

Decay chains, 7
   differential equations, 8, 9
Delayed neutrons, 70
(d,n) reaction, 69
D_2O
   as reactor moderator, 43, 44, 51
   diffusion length, 43
   Fermi age, 42
   moderating ratio, 42
   slowing-down length, 43, 51
Dose, 124
   calculation, 124, 130, 131
   in mixed radiation field, 125
   limit of detection, 125, 126
   measurement, 125
   per unit fluence, 124
Dosimeter, 131
   activation detector as, 131, 132, 133
^{155}Dy, 25
^{167}Dy, 26
^{165}Dy, 25, 26, 64
^{164}Dy, 64
^{159}Dy, 25, 26
^{157}Dy, 25

E

^{161}Er, 26
^{163}Er, 26
^{165}Er, 26
^{167}Er, 26
^{169}Er, 26
^{171}Er, 26
^{173}Er, 26
^{150}Eu, 25
^{152}Eu, 25, 156
^{154}Eu, 25
^{155}Eu, 24, 25
^{156}Eu, 25
^{157}Eu, 25
^{158}Eu, 25
^{160}Eu, 25

F

^{18}F, 13, 68, 69, 117, 121, 123, 126, 128, 129, 172, 188, 189
^{19}F, 55, 56, 69, 75
^{20}F, 13, 55, 56, 74, 75
^{21}F, 13
^{22}F, 13
Fast neutron activation, 10, 60, 61
   calculation of yield, 57
Fast neutron flux, 49, 50, 53, 54
   measurement, 53
Fast neutrons, 10
   flux, 49, 50, 53, 54
   production, 10
^{53}Fe, 15
^{55}Fe, 15
^{56}Fe, 55
^{58}Fe, 189
^{59}Fe, 15, 156, 157, 159, 171, 189
^{61}Fe, 15
Fermi age, 42
   for Be, 42
   for C, 42
   for D_2O, 42
   for H_2O, 42
Fission spectrum, 39
Fluence, 115, 118, 124, 125, 127, 129, 130
   analysis, 127, 128, 129, 130
Fluence assessment, 115, 118, 126, 127, 129, 133
   choice of monitor reaction, 119
   choice of target, 119
   dose calculation, 124 130, 131
   in a mixed beam, 127, 133
   neutron, 121, 122, 126, 128, 129
   π-meson, 127, 130
   proton, 120, 121, 126, 127, 128, 129
   with activation detector, 118, 125, 127, 133, 134

G

(γ,γ') reaction, 31
(γ,n) reaction, 31
(γ,p) reaction, 31
γ-radiation, 39, 45, 116, 188, 190
γ-ray energy
   assessment, 146

SUBJECT INDEX 217

γ-ray spectrometry, 33, 85, 86, 176
   by Ge-Li semiconductor detector, 146, 147, 148, 149, 150, 151, 152, 176, 177
   by scintillator detector, 148, 149, 150, 151, 176
   drift control, 93, 94, 147
   γ-emission rate, 149.
   γ-emitter identification, 148
   γ-ray energy determination, 146, 147
   high-resolution, 137
   instrumental, 85
   least squares method, 98
   limitations, 33, 34, 35
   linear superposition, 97
   matrix method, 98
   noise control, 92
   photopeak counting rate, 148, 149
   resolution, 96, 97
   spectrum stripping, 97
^{68}Ga, 16
^{70}Ga, 15, 16
^{72}Ga, 15, 16
^{73}Ga, 16
^{74}Ga, 16
^{151}Gd, 25
^{153}Gd, 25, 147
^{155}Gd, 71
^{156}Gd, 71
^{159}Gd, 25
^{161}Gd, 25
^{71}Ge, 16
^{75}Ge, 16, 75
^{77}Ge, 16
^{79}Ge, 16
Ge-Li drifted detector, 139, 143, 144, 145, 146, 149, 150
   amplifiers, 145, 146
   analyser, 146
   coaxial
   construction, 141, 142, 143
   efficiency, 151, 152
   encapsulated, 143
   function, 139, 140, 141, 144
   in activation analysis, 152, 155-160
   in analysis of high-purity materials, 155
   in γ-ray spectrometry, 146, 147, 148, 149, 150, 151
   in biological analyses, 157-160
   in geochemistry, 155-7
   non-encapsulated, 143
   planar, 143
   resolution, 141, 142, 144, 145
   sensitivity, 152, 153, 154, 155
Graphite
   as reactor moderator, 42, 51
   diffusion length, 43
   Fermi age, 42
   Slowing-down length, 43, 51

H

^{1}H, 189, 198
   cross-section, 51
   as reactor moderator, 41
^{2}H, 45, 62, 189, 198
   as reactor moderator, 41
^{3}H, 13
^{4}H, 13
^{6}He, 13
^{7}He, 13
^{173}Hf, 27
^{175}Hf, 27
^{178}Hf, 26, 27
^{179}Hf, 26, 27
^{180}Hf, 26, 27
^{181}Hf, 26, 27, 155, 156
^{183}Hf, 27
^{195}Hg, 29
^{196}Hg, 171, 189
^{197}Hg, 28, 29, 189
^{199}Hg, 28, 29
^{203}Hg, 28, 29, 93, 147, 150, 171
^{205}Hg, 28, 29
High energy radiation, 116
   cross section, 117
   source, 116
   spectrum, 116, 118
^{161}Ho, 26
^{162}Ho, 26
^{163}Ho, 26
^{164}Ho, 26
^{165}Ho, 64
^{166}Ho, 26
^{167}Ho, 26
^{168}Ho, 26
^{170}Ho, 26
H_2O
   advantages as moderator, 43, 51

H_2O (contd.)
    as reactor moderator, 41, 51
    diffusion length, 43
    Fermi age, 42
    moderating ratio, 42
    slowing-down length, 43, 51
Human body
    elemental composition, 165

I

^{123}I, 23
^{124}I, 23
^{125}I, 23, 184, 192
^{126}I, 22, 23
^{127}I, 190, 194, 199
^{128}I, 22, 23, 160, 173, 190, 199
^{129}I, 22, 23, 189, 190, 191, 192, 199
^{130}I, 23, 189, 190, 199
^{131}I, 22, 23
^{132}I, 23
^{133}I, 23
^{134}I, 23
^{135}I, 48
^{136}I, 23
^{111}In, 21
^{112}In, 20, 21
^{113}In, 21, 53
^{114}In, 20, 21
^{115}In, 20, 21, 53
^{116}In, 20, 21, 53
^{117}In, 20, 21
^{118}In, 21, 22
^{119}In, 21
^{120}In, 21
^{121}In, 21
^{122}In, 21
^{124}In, 21
Induced activity, 6
Interference, 7, 11, 29
^{189}Ir, 28
^{190}Ir, 28
^{191}Ir, 27, 28
^{192}Ir, 28
^{194}Ir, 28
^{195}Ir, 28
^{196}Ir, 28
^{198}Ir, 28

K

^{38}K, 14
^{40}K, 14, 196, 197
^{41}K, 57, 189
^{42}K, 14, 112, 158, 159, 189
^{43}K, 14
^{44}K, 14
^{45}K, 14
^{46}K, 14
^{48}K, 14
^{77}Kr, 17
^{79}Kr, 17
^{81}Kr, 17
^{83}Kr, 16, 17
^{85}Kr, 17
^{87}Kr, 17

L

^{135}La, 24
^{136}La, 24
^{137}La, 24
^{140}La, 24, 156, 157
^{142}La, 24
^{5}Li, 13
^{6}Li, 42, 62, 69, 76
^{7}Li, 42, 75, 76
^{8}Li, 13, 75, 76
Linear accelerator, 86, 172
    in clinical activation analysis, 172
Liver
    elemental composition, 165
^{173}Lu, 27
^{174}Lu, 26, 27
^{176}Lu, 26, 27
^{177}Lu, 26, 27, 147
^{178}Lu, 27
^{179}Lu, 27
^{180}Lu, 27

M

Mass spectrometry, 35
Mercury
    assay, 32
Meson, 116, 117, 118, 125, 127
^{23}Mg, 13
^{25}Mg, 75
^{26}Mg, 189, 194
^{27}Mg, 13, 55, 173, 189
^{53}Mn, 15
^{54}Mn, 15, 85, 86, 142
^{55}Mn, 52
^{56}Mn, 15, 52, 55, 158
^{57}Mn, 15

^{58}Mn, 155
^{91}Mo, 18
^{93}Mo, 18, 19
^{99}Mo, 18, 19
^{101}Mo, 18, 19
Moderating ratio, 42
  of Be, 42
  of C, 42
  of D_2O, 42
  of H_2O, 42
Moderator, 40, 41, 42
  for ^{235}U fission, 40, 41, 42
  ideal, 40
  mode of operation, 41
  moderating ratio, 42
  properties, 40, 41

N

^{13}N, 13, 45, 60, 66, 69, 189, 194, 196, 197, 198
^{14}N, 60, 69, 196, 198
^{15}N, 60
^{16}N, 13, 45, 53, 60, 66, 75
^{17}N, 13
^{18}N, 13
^{22}Na, 13, 142, 157
^{23}Na, 55, 75
^{24}Na, 13, 52, 119, 121, 123, 126, 128, 146, 147, 153, 154, 159, 171, 173, 188, 194, 196, 197, 202
^{25}Na, 75
  as flux monitor, 52
  γ-ray energy, 52
^{23}Na, 55
^{24}Na, 112, 129, 150, 158, 172
^{25}Na, 13
^{26}Na, 13
(n,α) reaction, 53, 54, 56, 57, 60, 62, 71, 75, 125
^{91}Nb, 18
^{92}Nb, 18
^{93}Nb, 18
^{94}Nb, 18
^{95}Nb, 18, 157
^{96}Nb, 18
^{97}Nb, 18
^{98}Nb, 18
^{100}Nb, 18
^{141}Nd, 24, 25
^{144}Nd, 25
^{147}Nd, 24, 25, 156
^{149}Nd, 24, 25, 48
^{151}Nd, 24, 25
^{19}Ne, 13
^{20}Ne, 56
^{21}Ne, 172
^{23}Ne, 13, 75
Neutron absorption determination, 71
Neutron activation, 4, 56, 58, 60
  irradiation period, 64, 65, 66
  recoil protons, 68
Neutron activation analysis, 4, 5, 6, 54, 63, 73, 76, 77, 78
  by fast neutrons, 10, 60
  by thermal neutrons, 54
  data processing by computer, 76, 77
  limits of detection, 59, 61, 62
  sample irradiation, 62, 63, 64, 66
  use of reactor pulses, 73
Neutron diffraction, 71
Neutron energy, 39, 56
  epithermal region, 52
  spectrum, 39
Neutron flux, 4, 5, 6, 37, 48, 49, 50, 51, 52, 53
  energies, 39, 49, 50, 51, 56
  fast, 49, 50, 53, 54
  measurement, 51, 52, 53, 54
  ratio of thermal to fast, 50, 51
  thermal, 49, 50, 51, 52, 53, 58
  variation through reactor, 50
Neutron generator, 86, 171
  in clinical activation analysis, 171
Neutron radiography, 71
Neutrons
  fast, 5, 10
  resonance activation, 52
  source, 4, 10, 37
  thermal, 4, 37
(n, γ) reactions, 5, 11, 29, 51, 52, 53, 54, 55, 56, 57, 58, 60, 62, 71, 115, 146, 148
  irradiation products, 5
^{57}Ni, 15
^{59}Ni, 15
^{63}Ni, 15
^{64}Ni, 189
^{65}Ni, 15, 189

^{67}Ni, 15
(n,n') reaction, 11, 53, 57, 58, 71, 74, 75
$(n, 2n)$ reaction, 11, 57, 58, 62, 69, 74, 75, 125
(n, p) reaction 11, 53, 54, 56, 57, 60, 62, 69, 75, 117, 125
Nuclear species
    production, 8
    disappearance, 8

O

^{15}O, 13, 45, 66
^{16}O, 53, 60, 69, 75
^{18}O, 60, 68, 69, 172, 188, 189
^{19}O, 13, 45, 60, 66, 75
^{183}Os, 27, 28
^{185}Os, 27, 28
^{189}Os, 27, 28
^{190}Os, 27, 28
^{191}Os, 27, 28
^{193}Os, 27, 28
^{195}Os, 28

P

^{30}P, 13
^{31}P, 11, 54, 56, 121, 122, 123, 196
^{32}P, 13, 14, 53, 54, 56, 121, 123, 128, 130, 171, 194, 198
^{33}P, 13
^{34}P, 13, 14, 75
^{36}P, 13
^{233}Pa, 156
^{203}Pb, 29, 189
^{204}Pb, 189
^{205}Pb, 29
^{206}Pb, 75
^{207}Pb, 29, 75
^{208}Pb, 75
^{209}Pb, 29
^{212}Pb, 150
^{101}Pd, 19
^{103}Pd, 19, 20
^{107}Pd, 19, 20
^{109}Pd, 19, 20
^{111}Pd, 19, 20
^{113}Pd, 20
(p, γ) reactions, 68
^{143}Pm, 25
^{144}Pm, 25
^{145}Pm, 25
^{147}Pm, 24, 25
^{148}Pm, 25
^{149}Pm, 24, 25
^{150}Pm, 25
^{151}Pm, 24, 25
^{152}Pm, 25
^{154}Pm, 25
(p,α) reaction, 69
(p,n) reaction, 68, 69, 117
(p, pn) reaction, 117, 125, 132
^{140}Pr, 24
^{142}Pr, 24
^{143}Pr, 24
^{144}Pr, 24
^{145}Pr, 24
^{146}Pr, 24
^{147}Pr, 24
^{148}Pr, 24
^{150}Pr, 24
Prompt γ-rays, 70
    in analysis, 70, 71
^{189}Pt, 28
^{191}Pt, 28
^{193}Pt, 28, 29
^{195}Pt, 28, 29
^{197}Pt, 28, 29
^{199}Pt, 28, 29
^{201}Pt, 29
^{239}Pu, 38, 48, 70, 93
Pulsed reactor, 6, 72
    effect on activity, 73
    in neutron activation analysis, 72, 73, 74, 75, 76

R

^{226}Ra, 142
Radiation, 115
    hazard, 115
Radioactivity
    assay, 31, 32
Radiochemical separation, 32, 77, 81, 82, 85
    automation, 77
Radionuclide, 12-29, 34, 118
    decay, 34
    from ^{235}U fission
    half-life, 12-29
    saturation-activity, 12-29
^{83}Rb, 17
^{84}Rb, 17
^{86}Rb, 17, 18, 156, 157, 159, 173

^{88}Rb, 17
^{180}Re, 28
^{183}Re, 28
^{184}Re, 27, 28
^{186}Re, 27
^{188}Re, 27, 28
^{189}Re, 28
^{190}Re, 27, 28
^{192}Re, 27
Reactors, 4, 37, 86, 88
   ATPR, 72
   cooling, 45
   construction, 49, 50
   control, 46
   cost, 44
   critical mass, 46
   design, 44
   fission products, 38
   fuel elements, 47
   fuel-element replacement, 48
   in activation analysis, 37, 171, 195
   MTR, 47
   operation, 38, 73, 88
   pool-type, 44, 47, 49, 171
   pulsing, 72
   samarium poisoning, 48
   shielding, 44
   steady state, 46
   TRIGA, 44, 45, 47, 49, 50, 54, 57, 58, 63, 65, 66, 67, 68, 72, 73, 76, 87
   water-boiler type, 46
   xenon poisoning, 47
Recoil protons—
   in determination of N, 69
   in determination of O, 69
   in neutron activation, 68, 69
   interference, 69
^{101}Rh, 19
^{102}Rh, 19
^{103}Rh, 19, 20
^{104}Rh, 19, 20
^{105}Rh, 19
^{106}Rh, 19, 20
^{107}Rh, 19
^{108}Rh, 19
^{110}Rh, 19
^{95}Ru, 19
^{97}Ru, 19
^{103}Ru, 19, 157
^{105}Ru, 19

S

^{31}S, 53
^{32}S, 53, 54, 56, 121, 122, 123, 128, 130
^{33}S, 53, 121
^{34}S, 75, 121
^{35}S, 14
^{36}S, 121, 189
^{37}S, 14, 189, 194
^{119}Sb, 22
^{120}Sb, 22
^{122}Sb, 21, 22
^{124}Sb, 21, 22, 156, 157
^{125}Sb, 21, 22
^{126}Sb, 22
^{127}Sb, 22
^{128}Sb, 22
^{130}Sb, 22
^{44}Sc, 14, 172
^{46}Sc, 14, 74, 75, 150, 156, 159
^{47}Sc, 14
^{48}Sc, 14, 172, 189
^{49}Sc, 14
^{50}Sc, 14
Scintillation detector, 33, 89, 148, 149
   efficiency, 151
   sensitivity, 152, 153, 154, 155
^{74}Se, 58, 189
^{75}Se, 16, 17, 58, 189
^{76}Se, 58
^{77}Se, 16, 17, 58, 74, 177
^{78}Se, 58
^{79}Se, 16, 17
^{81}Se, 16, 17
^{83}Se, 16, 17
Semiconductor detector, 137, 138, 139
   as γ-ray detector, 137
   Ge-Li drifted, 139-146
^{27}Si, 13
^{28}Si, 11
^{29}Si, 53
^{30}Si, 55
^{31}Si, 13, 121, 123
^{33}Si, 13
^{35}Si, 13
^{37}Si, 13
^{143}Sm, 25
^{145}Sm, 24, 25
^{149}Sm, 48
   reactor poison, 48
^{150}Sm, 48

^{151}Sm, 24, 25
^{153}Sm, 24, 25
^{155}Sm, 24, 25
^{157}Sm, 25
^{111}Sn, 21
^{113}Sn, 21
^{117}Sn, 20, 21
^{119}Sn, 21, 22
^{121}Sn, 21, 22
^{123}Sn, 21, 22
^{125}Sn, 21, 22
^{127}Sn, 22
Soil
    elemental composition, 165
Spallation reaction, 117, 123, 126, 128
Spectrophotometry, 35
^{83}Sr, 17
^{84}Sr, 189
^{85}Sr, 17, 150, 157, 159, 189
^{87}Sr, 17, 18, 160
^{89}Sr, 17, 18
^{91}Sr, 18,
^{93}Sr, 18
Standardization, 32
    difficulties, 32
    internal standard method, 32

T

^{51}T, 15
^{179}Ta, 27
^{180}Ta, 27
^{182}Ta, 27, 156
^{183}Ta, 27
^{184}Ta, 27
^{186}Ta, 27
Target, 116
^{149}Tb, 117, 121, 123, 126
^{155}Tb, 25
^{156}Tb, 25
^{157}Tb, 25, 26
^{158}Tb, 25, 26, 75
^{159}Tb, 75
^{160}Tb, 25
^{161}Tb, 25
^{162}Tb, 25, 26
^{163}Tb, 25
^{164}Tb, 25
^{95}Tc, 19
^{96}Tc, 19
^{97}Tc, 19
^{98}Tc, 19
^{99}Tc, 18, 19
^{100}Tc, 19
^{101}Tc, 18, 19
^{102}Tc, 19
^{104}Tc, 19
^{119}Te, 22
^{121}Te, 22, 23
^{123}Te, 22, 23
^{125}Te, 21, 22, 23
^{127}Te, 22, 23
^{129}Te, 22, 23
^{131}Te, 22, 23
^{133}Te, 23
^{135}Te, 47
^{228}Th, 150, 151, 152
Thermal column, 51
Thermal neutron activation, 7, 54
    calculation of yield, 57
    conditions, 7
    interfering reactions, 7, 55
Thermal neutron flux, 49, 50, 51, 52, 53, 58
    measurement, 51, 52, 53
Thermal neutrons, 4
    diffusion length, 43
    flux, 49, 50, 51, 52, 53
    flux monitor, 52
    migration length, 43
    production, 4
    slowing-down length, 42
^{45}Ti, 14
^{50}Ti, 57, 58
^{51}Ti, 14, 55, 57, 58
^{55}Ti, 58
^{202}Tl, 29
^{204}Tl, 29
^{206}Tl, 29
^{207}Tl, 29
^{208}Tl, 29, 150
^{167}Tm, 26
^{168}Tm, 26
^{170}Tm, 26
^{171}Tm, 26
^{172}Tm, 26
^{173}Tm, 26
^{174}Tm, 26
^{176}Tm, 26
(t,n) reaction, 62, 69

SUBJECT INDEX

U

^{233}U, 38
^{234}U, 44
^{235}U, 37, 38, 44, 45, 46, 47, 48, 50, 62, 70, 72
    cross-section, 39
^{236}U, 38
^{238}U, 41, 44, 48, 50, 70

V

^{49}V, 14, 15
^{51}V, 55, 57
^{52}V, 14, 15, 75
^{53}V, 15
^{54}V, 15
Van de Graaff accelerator, 86

W

^{179}W, 27
^{181}W, 27, 28
^{183}W, 27, 74
^{185}W, 27, 28
^{187}W, 27, 28
^{189}W, 27

X

^{123}Xe, 23
^{125}Xe, 22, 23
^{127}Xe, 22, 23, 24
^{129}Xe, 23
^{131}Xe, 22, 23, 24
^{133}Xe, 23
^{135}Xe, 22, 23, 47
    (n,γ) reaction, 47, 48
    reactor poison, 47
^{137}Xe, 22, 23
X-rays, 116

Y

^{88}Y, 18
^{89}Y, 75
^{90}Y, 17, 18
^{91}Y, 18
^{92}Y, 18
^{93}Y, 18
^{94}Y, 18
^{96}Y, 18
^{167}Yb, 26
^{169}Yb, 26
^{175}Yb, 26, 27
^{177}Yb, 26, 27

Z

Z value, 30, 41, 52, 60
    in choice of flux monitor element, 51, 52
    in choice of moderator, 41
^{63}Zn, 15
^{65}Zn, 15, 154, 155, 157, 159, 171, 173
^{69}Zn, 15, 16
^{70}Zn, 189
^{71}Zn, 15, 16, 189
^{73}Zn, 16
^{89}Zr, 18
^{90}Zr, 18
^{93}Zr, 18
^{95}Zr, 18, 157
^{97}Zr, 18